POLITICAL PHILOSOPHY NOW

Chief Editor of the Series:
Howard Williams, Emeritus Professor, Aberystwyth University, Wales; Honorary Distinguished Professor, Cardiff University, Wales

Associate Editors:
Wolfgang Kersting, University of Kiel, Germany
Renato Cristi, Wilfrid Laurier University, Waterloo, Canada
Susan Meld Shell, Boston College, Massachusetts, USA
David Boucher, Cardiff University, Wales

Affiliate Editors:
Peter P. Nicholson, formerly of University of York, England
Steven B. Smith, Yale University, USA

Political Philosophy Now is a series which deals with authors, topics and periods in political philosophy from the perspective of their relevance to current debates. The series presents a spread of subjects and points of view from various traditions which include European and New World debates in political philosophy.

Also in series

Kant's Critique of Practical Reason:
A Philosophy of Freedom
Otfried Höffe

How Kant Matters for Biology:
A Philosophical History
Andrew Jones

The Moral Standing of the State in
International Politics: A Kantian Account
Milla Emilia Vaha

Kant's Doctrine of Right *in the Twenty-first Century*
Edited by Larry Krasnoff, Nuria Sánchez Madrid and Paula Satne

Kant's Political Legacy: Human Rights, Peace, Progress
Luigi Caranti

Nietzsche On Theognis of Megara
Renato Cristi and Oscar Velásquez

Nietzsche and Napoleon: The Dionysian Conspiracy
Don Dombowsky

Politics and Teleology in Kant
Edited by Paul Formosa, Avery Goldman and Tatiana Patrone

Identity, Politics and the Novel:
The Aesthetic Moment
Ian Fraser

Kant on Sublimity and Morality
Joshua Rayman

Politics and Metaphysics in Kant
Edited by Sorin Baiasu, Sami Pihlstrom and Howard Williams

Hegel and Marx: After the Fall of Communism
David MacGregor

Francis Fukuyama and the end of history
Howard Williams, David Sullivan and E. Gwynn Matthews

POLITICAL PHILOSOPHY NOW

The Ethics of Remote Warfare

Lily Hamourtziadou

UNIVERSITY OF WALES PRESS • 2024

© Lily Hamourtziadou, 2024

All rights reserved. No part of this book may be reproduced in any material form (including photocopying or storing it in any medium by electronic means and whether or not transiently or incidentally to some other use of this publication) without the written permission of the copyright owner except in accordance with the provisions of the Copyright, Designs and Patents Act. Applications for the copyright owner's written permission to reproduce any part of this publication should be addressed to the University of Wales Press, University Registry, King Edward VII Avenue, Cardiff CF10 3NS.

www.uwp.co.uk

British Library Cataloguing-in-Publication Data
A catalogue record for this book is available from the British Library.

ISBN 978-1-83772-162-7
e-ISBN 978-1-83772-163-4

The right of Lily Hamourtziadou to be identified as author of this work has been asserted in accordance with sections 77, 78 and 79 of the Copyright, Designs and Patents Act 1988.

Typeset by Marie Doherty
Printed by CPI Antony Rowe, Melksham, United Kingdom

Contents

'Rules-based Orders' by *Dr Barry J. Ryan*		vii
List of Tables		ix
Foreword by *Sorin Baiasu*		xi
Introduction: Twenty-First Century Wars and Technology		1
1	Three Approaches to Ethics	11
2	Killing from Afar: The Terror of the Air	33
3	From the Bomber to the Drone	57
4	Remote Killing in the War on Terror	93
5	Remote Killing and the War in Ukraine	135
Conclusion: Remote Warfare and the (New) Ethics of War		167
Notes		177
Select Bibliography		201
Index		209

Rules-based Orders

    wan

    ton

soup,

    a mayhem

    for starter, then

    spaghetti boys

            bombing Belgrade,

                      Kabul

    is

    kidney pie,

    garnished by

            Helmand

    Mayonnaise.

    Dessert

        is in the desert:

    Toad in Hole.

            Hanged Saddam

    Or pig on a spit

        From Sirte.

→

We eat the east.

    This is what we live on

    Until

that   time

    but it stops.

List of Tables

Table 4.1	Iraq Body Count Incident x001	99
Table 4.2	Iraq Body Count Incident x002	99
Table 4.3	Iraq Body Count Incident x003	99
Table 4.4	Iraq Body Count Incident x038	100
Table 4.5	Iraq Body Count Incident a6384	100
Table 4.6	Monthly civilian deaths 2003–13	103
Table 4.7	The ISIS years (monthly civilian deaths 2014–17)	105
Table 4.8	Iraq Body Count Incident a5888	106
Table 4.9	Iraq Body Count Incident a5885	107
Table 4.10	Iraq Body Count Incident a5878a	108
Table 4.11	Iraq Body Count Incident a6317	108
Table 4.12	Iraq Body Count Incident a6262	123
Table 4.13	Massacres and mass graves by ISIS	131

Foreword

We avoid civil wars by a complex system of powers, which aim to prevent and resolve conflicts through an intricate mechanism: we have laws, for the observance of which various incentives are provided (such as the threat of punishment), procedures for adjudication in cases of disputes (for instance, through trials in a court), and processes of enforcing such decisions (including the prison system, but also systems of fines and penalties for non-criminal offences). Conceivably, we could avoid wars between states with the help of a similar system, this time at international level. But there are well known problems concerning the risks of having a world state or the difficulties of establishing an international court with jurisdiction over inter-state conflicts or the justification of a legitimate executive power for the enforcement of decisions that such courts would make.

Attempts to build such an international system of relations between states able to avoid war have been made and continue to be made. Various institutions are designed in theory and practice with the aim of reducing the likelihood of armed conflicts and resolving, in a fair way, international disputes. Although progress is never linear, clear steps forward have been made. Yet, the Kantian suggestion of replacing war with litigation relies on an assumption that is currently under threat: this is the assumption that it is possible to identify responsibility for actions and consequences, and that rules of imputation can offer clear answers to questions of merit, innocence and culpability. The assumption is, however, under pressure in the context of warfare.

The way that war is conducted is constantly changing. It is not only that weapons are increasingly destructive, they are also increasingly autonomous, capable of guiding themselves towards targets and even capable of selecting targets according to some specific criteria. Furthermore, it is increasingly easy to programme, activate and observe these weapons remotely creating the illusion that war is a computer game or a Hollywood movie. Moreover, war is no longer localised – consequences will occur in specific places, but they can be caused from any place with an internet connection and the right software. When the consequences of war do not seem real, responsibility for these consequences also seems an insignificant issue. What follows is a situation in which increasingly devastating consequences can be inflicted, but it is increasingly difficult to realise their magnitude and identify those responsible for them.

What can be done? For those like us, readers of this book, interested in the theoretical aspects of war, it is necessary to present clearly the new realities of armed conflict and the mechanisms through which the increasingly serious consequences of war are masked as lacking importance; moreover, the new ethical issues arising from the changing landscape of contemporary warfare need to be formulated and the challenge of finding a proper response has to be approached as decisively as possible. New avenues for research and discussion should be proposed to address crucial problems, such as that of ascertaining responsibility and judging imputability.

The Ethics of Remote Warfare is an important text taking up all these tasks. Written by an academic with experience of pioneering work on armed conflicts and their tragic consequences, the book responds to an urgent need for a clear and critical analysis of the new realities of war. The text examines some of the latest warfare practices, which evolved with the development of new technologies, and it focuses on some of the most worrying developments, such as the US War on Terror, and the war in Ukraine. This book focuses on a new set of ethical issues raised by the most recent developments in technology and geopolitics. For instance, it looks at the moral problem of AI-powered weapons, discusses the 'cleaner' and 'targeted' killings, reflects on the shifting of risk and responsibility to other agents, such as local allies or security contractors, and explores the ethics of drone usage or the masking of responsibility through the virtualisation of war. Its theoretical contribution is very significant both in the narrower context of remote warfare and in the broader setting of ethically grounded attempts to replace war with adjudication of conflicts through law.

Sorin Baiasu
Professor of Philosophy
Keele University
August 2023

Introduction: Twenty-First Century Wars and Technology

> And others are proud of their modicum of righteousness,
> and for the sake of it do violence to all things: so that
> the world is drowned in their unrighteousness.
>
> Ah! how ineptly cometh the word 'virtue' out of their mouth! And when they say: 'I am just,' it always soundeth like: 'I am just – revenged!'
>
> With their virtues they want to scratch out the eyes of their enemies; and they elevate themselves only that they may lower others.[1]

Changing parameters of conflict within a context of changing global power relations

Talking of war inevitably involves talking about it in terms of right and wrong, even when we argue that war lies outside moral judgement, that in war self-interest and necessity prevail over morality. Thus, military strategy – the other language of war – can override the ethics of war. Nevertheless, we say that a particular war is just or unjust, or that a war is being fought justly or unjustly.[2] The study of military ethics, especially just war theory, is increasingly institutionalised in programmes of military training. Resistance to war has also drawn on ethical practices and discourses of morality.

> Anti-war movements and human rights organisations often appeal to moral standards as a means of generating public opposition, seeking to constrain militaries and the state through moral scrutiny. Ethics has therefore become an increasingly important terrain on which war and military activity is understood, legitimated and contested.[3]

The parameters of military conflict in the twenty-first century, compared with the Cold War period, have changed as a result of technological progress and a changing architecture of global power relations. The past twenty years have seen a redistribution of economic power among the world's great powers. China, which was one-eighth the size of the US economy, is now roughly the same size in purchasing power terms. Other countries have also shown consistently high rates of growth – for example, India. On the other hand, we see a relative decline of the West. Since 2008, every major power has undergone some form of deep internal economic readjustment. In the military and technological domains, US predominance continues, but new technologies have empowered small groups and individuals, further eroding the state's monopoly on violence. The combination of economic crises, the social and political fragility that they have induced, and new technologies, has created geopolitical shifts. Rising geopolitical competition increasingly happens through proxies rather than direct confrontation, so the prospect of conventional war between major powers is low.

Managing and creating grey zone conflict – activities that occur between peace and war – has been an important part of great powers' international strategy. Many activities fall into this grey zone: from nefarious economic activities and cyberattacks, to mercenary operations, targeted assassinations, engaging surrogates in a conflict, instigating revolutions and disinformation campaigns. Grey zone activities combine non-military and quasi-military tools and are at the threshold of armed conflict. They aim to thwart, destabilise, weaken or attack an adversary. The onset of new technologies has provided states and non-state actors with more tools to operate and avoid clear categorisation, attribution, responsibility and detection, in a context of globalisation, as well as the proliferation of the internet and social media. Increasingly, nations seek to promote their national objectives through aggression conducted covertly, or with obscure attribution, morality and justification.

Twenty-first century technology, military and political strategy blur boundaries between peace and war, the military and the civilian, the conventional and the unconventional, the state and the non-state, the legitimate and the criminal, the moral, the immoral and the amoral. Cold War has given way to a 'Cool War' involving constant offensive measures that seek to damage or weaken rivals through violations of sovereignty and penetration of defences, using new technologies that change the paradigm of conflict. Actions are carried out by a long chain of performers, all giving and taking orders, everyone has a specific, focused task, while the people affected do not appear fully human. 'Modernity did not make people more cruel', Bauman wrote, 'it only invented a way in which cruel things could be done by non-cruel people.'[4]

A key feature of many of the vital systems introduced for the digital age, including internet providers, search engines, hardware manufacturers and software developers, was that they were owned and operated by private companies with global interests, wrote Freedman in *The Future of War*.[5] Smartphones carry capabilities like satellite imaging, navigations, data stores, instant, encrypted communications. These readily accessible systems made it possible for individuals and small groups to hurt others, as communities were exposed to new risks in hybrid wars. Hybrid warfare, often involving the use of drones, makes war elusive, ambiguous, easier, cheaper and less risky. Hybrid attacks are marked by vagueness, complicating attribution. When violence is committed remotely, who is responsible or complicit? Everybody, anybody or nobody?

In March 2022, just days after Russia invaded Ukraine, the White House announced that its military support package to Ukraine included 100 tactical unmanned aerial systems, the Switchblade loitering munition known as the 'kamikaze drone', which had gained notoriety in the 2020 Nagorno-Karabakh war. The United States was committed to providing over 700 Switchblades in total. Once airborne, loitering munitions hunt for a target and crash into it. Years earlier, Russia had created its own loitering munitions to counter insurgencies in Syria:

> Notably, the Russian manufacturer Zala-Aero Group designed the Lancet to autonomously locate and strike targets in designated areas, raising morality issues with the human-out-of-the-loop approach. This is the only system in this conflict that relies on autonomous hunting and killing, a morally gray area for many in the defence industry.[6]

Debates about the rules and ethics of war have taken place for thousands of years, but those debates have relevance and new applications for theorists and practitioners today, as technology enables more and more ways to fight enemies: drones, satellites, computers, global positioning systems (GPS). The use of drones has been declared ethical and safe, effective in killing and enabling the West to fight 'clean wars' for over two decades. As the use of drones has become much wider, employed by governments and militaries in civil, defensive and aggressive wars, and by superpowers and small states alike, is war becoming cleaner, moral and just, or is warfare reduced to an endless series of cold-blooded murders?

Increasingly virtual wars are regarded as virtuous wars, as there are increasing efforts 'to effect ethical change through technological and martial means'.[7] While by 2022 such means have become widespread, it is the United States that has led this virtual revolution. Its diplomatic and military policies are based largely on technological and representational forms of deterrence and compulsion that could be described as 'virtuous war', at the heart of which is 'the technical capability and ethical imperative to threaten and, if necessary, actualise violence from a distance "with no or minimal casualties"'.[8] Virtual/virtuous war gives the digitally advanced a strategic advantage, while promoting a vision of bloodless, humanitarian and hygienic warfare. Unlike other forms of warfare, virtual/virtuous war can 'commute death', keeping it out of sight and out of mind. Herein, warns Der Derian, lies the most morally dubious danger:

> In simulated preparations and virtual executions of war, there is a high risk that one learns how to kill but not to take responsibility for it. One experiences 'death' but no the tragic consequences of it. In virtuous war we now face not just the confusion, but the pixilation of war and game on the same screen.[9]

Is virtualisation the continuation of war by other means? Is virtuality replacing the reality of war?

The Ethics of Remote Warfare addresses ethical dilemmas in war arising from technological progress: air attacks and drone warfare. Have we found new, moral, ways of fighting wars? Or produced 'sanitised' versions of the same wars? Or is morality not applicable to war? Moreover, how do the new types of warfare fit into our democratic political systems and to what extent do they undermine human security and violate human rights?

As the twentieth century was ending and the twenty-first century beginning, critical geopolitical texts were infused with the vocabularies of normative moral judgement. Dowler and Sharp wrote of a moral proximity and anger 'at injustice, exploitation and subjugation' in an critical anti-geopolitical perspective that 'wants to see change'.[10] Dodds argued that there should be 'a sense of humanity and justice' and that 'critical geopolitics needs to continue to make a difference through our intellectual commitments and normative engagements'.[11] Drone warfare looms largest in the developing sphere of remote warfare, within which lies another fast-developing area: cyber conflict. Although far less lethal than killer drones, cyber weapons have also been deployed under rules of engagement that are unclear. Drone and cyber programs raise similar moral and legal issues that political leaders and publics will have to grapple with. Conducting attacks with drones or malware poses political and strategic challenges, as much as moral and legal ones.

This book explores the ethics of warfare in the context of an increasingly remote way of fighting enemies, by examining the advantages and the ethics of remote killing through engagement with three schools of thought on the morality of military force: realism, militarism and just war theory. Two case studies – the US War on Terror and the Russia-Ukraine war – will be used to illustrate and evaluate the ethical dilemmas and claims regarding the less deadly and more ethical nature of remote warfare.

Chapter 1 explores war and ethics by adopting three approaches to ethics in war: realism, militarism and just war theory. According to the realist approach, international relations are not amenable to moral determination in an anarchic and naturally bellicose world. The balance of power involves forces in constant flux and demands continual adjustment to changing circumstances, free of the constraints imposed by morality. In its purest form, it rejects the subjection of politics to ethics. As war is international relations *in extremis*, morality is thought of as a poor guide; the decision to go to war and any decisions about how to fight wars should be dictated not by moral sentiment but by pragmatic considerations of power and interest. For the realist, trying to subject war to moral limitations risks the delicate balance of the (imperfect) international order. The state has duties only to itself, but moral factors (for some realists) may come into play in war and its build up. Militarism seeks uncompromising victory and the annihilation of adversaries. It demands unconditional surrender or total submission. War is not a necessary evil, but a fundamental duty, or even a divine mission. It ceases to be instrumental and instead has its own intrinsic value, its own redemptive and transformative power. The just war tradition upholds the moral limitations of war. The just war theorist insists on the moral determination of war, where that is possible, and also affirm the moral primacy of peace over war. The moral determination of war is divided into the question of recourse to war and the subsequent conduct of war. The just war can never be total. War is fully intelligible as a human, and therefore moral, activity, informed by values and moral principles: a just war can only be waged as a last resort; a war is just only if it is waged by a legitimate authority; a just war can only be fought to redress a wrong suffered. Self-defence against an armed attack is always considered to be a just cause; war can only be just if it is fought with a reasonable chance of success; the ultimate goal of a just war is to re-establish peace; the violence used in the war must be proportional to the injury suffered; the weapons used in war must discriminate between combatants and non-combatants.

Chapter 2 examines remote killing through the development of air warfare. Control of the air has been important since the Second World War. It is argued that without control of the airspace, the success of any operation in the air and on the ground cannot be guaranteed. Air forces can range across the entire globe, giving air power one of its main characteristics: reach. Aircraft can engage the enemy at depth, through strikes against combatants, by attacking war industries or communications infrastructure. Strategic bombardment can win wars without surface forces. Aircraft are fast, flexible and can carry out attacks that are both deadly and terrifying; they can inflict mass killings and produce, to use a modern term, 'shock and awe' on the ground, maximising the psychological and material impact on the enemy, while minimising them for those conducting the airstrikes. Chapter 2 looks at the bombing campaigns of the Blitz, Dresden, Vietnam, Hiroshima and Nagasaki. The gas chambers of the Second World War, designed for the swift and mechanised extermination of millions of men, women and children, while sparing the killers the trauma of the mass murder of civilians, provide another example of distance killing. The strategic advantages and ethics of remote killing, in terms of actual and psychological distance, are examined through the lenses of realism, militarism and just war, and raise further questions: whose security must take priority? Are 'cleaner' killings more ethical?

Chapter 3 traces the emergence of drones, from spies to killers: the context of their emergence, their purposes, advantages and purported effectiveness, as well as their legality and 'ethical use'. It examines claims of 'precision bombing' and ethical concerns around the use of killer technology. In the process, it considers concepts of human rights, protected persons, just war and the consequences of such warfare for civilians. Much of the focus on the use of drones stems from the policies and tactics developed by Israel and the United States for 'targeted killing': seeking out of specific individuals for their role in military or terrorist operations and launching a precision military attack on that person. This practice has led to a great deal of public attention and scholarly debate over the use of drones. According to most moral theories, the use of violent and lethal force is permissible (if at all) under certain exceptional circumstances, such as in self-defence, the defence of another person, or to achieve a higher moral good or humanitarian benefit. How can we view the use of killer drones through the lenses of realism, militarism and just war? Is drone killing (of combatants or non-combatants) compatible with just war and international law?

Chapter 4 addresses remote killing in the US War on Terror, by examining the use of drones by a global hegemon, in low-level, persistent, remote and evasive modes of fighting. The United States' hybrid three-dimensional (3-D) warfare – delegation (shifting the burden of risk and responsibility on to other proxy actors, local allies, security contractors and irregular militias); danger-proofing (applying force, while minimising physical harm to US personnel, through the use of drones); and darkness (covert action and special forces operations, but also offensive cyber warfare) – is evaluated in terms of legal and moral impact through realism, militarism and just war. Does grey conflict mean grey ethics, in the political web of war? Are even targeted individuals without rights? The ethics regarding the United States' victims in the Middle East are at the heart of this case study, which focuses not only on the weapons and the perpetrators, but also on the dead, because no discussion on ethics, on what is morally (un)acceptable, can leave out the dead. Especially in the context of actions led by interests and strategies centred on exploiting and eliminating threats to security gains. The Iraqi dead, in particular, as documented by the researchers of the NGO Iraq Body Count, will form the basis on which to morally evaluate the hegemon's actions, during and following the invasion and occupation of Iraq, before and after the arrival of ISIS.

Chapter 5 looks at the use of drones by a smaller state, in self-defence, in the second case study: Ukraine. Both Ukraine and Russia used drones during the 2022–3 war, and Ukraine received from the United States hundreds of kamikaze Switchblade 300 and 600 drones, loitering munitions that can hang in the sky and smash down with fearful effect on their target. Smaller commercial drones have also provided tactical advantages, but is the use of drones more ethical, when the state in question is fighting a defensive war against a much larger, aggressive state? Chapter 5 examines the ethics surrounding a war of national defence, but also questions whether this is in fact self-defence or a remote war of a different kind – a proxy war – due to the NATO-US factor. A further ethical assessment follows, through the lens of the three perspectives, regarding the justification of the United States'-NATO's-Ukraine's war with Russia. The chapter ends by addressing the ethics of drone usage within the context of the profit and interest-driven global arms trade.

The Ethics of Remote Warfare concludes with reflections on the future of warfare and prospects for peace and justice. What is the impact of remote warfare on human security and peace? Will the effectiveness of the weapons lessen or increase their deployment? Do advances in technology make war more ethical, more likely, or do they change the nature of war so that it becomes more like an unsettled peace? Finally, in warfare without clear perpetrators, with an uncontrolled flow of arms, can there ever be accountability and justice? There is a very high probability that many if not all future wars will be 'drone wars', that drones will continue to play a defining role in future conflicts. When it comes to lethal autonomous weapons, we can foresee a future when warfare is dominated by intelligent, maybe even self-aware, machines, but we can also see terror-inspiring machines that can identify, track and target human beings of all ages and nationalities, without any human control and without any compliance with international humanitarian law or ethics. These machines, like any machines, may actually be incapable of following laws or ethics, thus, changing both the scale and the character of war. Moreover, the algorithmic targeting of humans strips them of their right to be recognised as fellow human beings, fellow moral agents, at the time of their death, while simultaneously making it very difficult to determine who or what is behind AI-powered weapons. New military technologies create new accountability gaps in armed conflict. Novel military technologies are also making more salient the accountability chasm that already exists in international humanitarian law: the relative lack of legal accountability for unintended 'awful but lawful' civilian harm.[12] Just-war-based international humanitarian law permits state military action, while attempting to minimise, rather than prevent, civilian harm. The law permits many acts that cause incidental civilian harm; AI warfare puts the law's accountability gap into sharper relief, highlighting the need for new accountability mechanisms, ones that reflect a sense of legal and of moral justice.

Drones may seem to offer a remedy for states and protection for individuals, but might in fact be among their greatest threats.

1 • Three Approaches to Ethics

Introduction

Even war has rules. Those rules that underpin international humanitarian law, developed in the twentieth century, can be found in the universally ratified Geneva Conventions 1949, and customary international law. The principles of international humanitarian law are universal and provide rules for the battlefield, sitting between legitimate military action and the objective of mitigating human harm, especially among civilians. The principle of humanity, for example, forbids the infliction of suffering, injury or death not necessary for the accomplishment of legitimate military purposes. During military operations, all efforts must be made to minimise any negative effects to the civilian population. All feasible precautions must be taken to avoid incidental loss of civilian life and damage to civilian objects. International humanitarian law has grown out of an international acceptance of the need for constraint in the waging and the conducting of war. Its fundamental rules reflect a universal humanitarian approach to warfare, and include the following:

- Persons *hors de combat* (outside of combat) are entitled to respect for their physical and moral integrity.
- It is forbidden to kill or injure an enemy who surrenders or who is *hors de combat*.
- The wounded and sick will be collected and cared for by the party to the conflict that has them in power. Protection also covers medical personnel, establishments, transports and *matériel*.
- Captured combatants and civilians under the authority of an adverse party are entitled to respect for their lives, dignity, personal rights and convictions.
- Everyone is entitled to benefit from fundamental judicial guarantees. No one will be subjected to physical or mental torture, corporal punishment or cruel or degrading treatment.

- Parties to a conflict and members of their armed forces do not have an unlimited choice of methods and means of warfare. It is prohibited to employ weapons or methods of warfare of a nature to cause unnecessary losses or excessive suffering.
- Parties to a conflict must at all times distinguish between the civilian population and combatants in order to spare civilian population and property.[1]

Connected to international humanitarian law, international criminal law is a body of law that defines international crimes, often taking place in the course of conflict, such as war crimes, genocide and the crime of aggression. In international criminal law we can see the influence of the famous Nuremberg Trials (1945–6), when leaders of Nazi Germany stood trial for war crimes and crimes against humanity. The trials established that humanity would be guarded by an international legal shield and that even heads of state would be held accountable and criminally responsible for aggression and crimes against humanity: 'The right of humanitarian intervention to put a stop to Crimes Against Humanity – even by a sovereign against his own citizens – gradually emerged from the Nuremberg principles affirmed by the United Nations.'[2]

The Nuremberg and the Tokyo tribunals lay the groundwork for modern international criminal law. The two world wars led the international community to pledge that 'never again' would such crimes be committed. Yet the horrors of the twentieth century did not stop, and indeed continued into the twenty-first century. Hundreds of thousands were killed in Vietnam, Sierra Leone, Iraq, Indonesia, East Timor, Somalia, Yugoslavia, Afghanistan and Rwanda. Few of the crimes committed in those conflicts were tried; in almost every case, those responsible for war crimes escaped punishment, which allowed the atrocities to continue. While international law was enforced and helped achieve some justice in the former Yugoslavia and in Rwanda, the international community has been reluctant to pursue justice for crimes committed during the War on Terror, despite the fact that an International Criminal Court was created as an essential component of a just world order under law, entering into force in July 2002.

The question of whether there is a place for justice and morality in warfare has often been asked. Are international law and international ethics essential to global politics, or is there little to no room for moral values and humanity when it comes to national security and the pursuit of political and economic interests? Although scholars 'have increasingly recognised the importance of moral values and ethical reasoning in international affairs, political realism continues to serve as the dominant paradigm in international relations', allowing 'little room for moral reasoning'.[3] But why should interests be amoral? Don't the pursuit of interests and foreign policy decisions involve moral judgement? Are not international relations based on important moral norms? Moreover, Amstutz argues, the development of the international community's rules, procedures and structures depend on values and norms that are partly moral. These norms provide the basis for moral claims on issues such as self-determination, human rights and the use of force.

This chapter examines three approaches to ethics in international relations and warfare, by scholars and theorists of alternative moral traditions of force: realism, militarism and just war. While querying the place of ethics in the pursuit of war and the conduct in war, questions arise regarding the implications, benefits and dangers of including and of excluding morality, when it comes to this extreme use of force.

Realism

Realists argue that international society is a realm of power, interests, the struggle for survival and necessity, not morality, humanity and justice, because no overarching authority exists to protect states and to resolve disputes. According to realism, international relations are not amenable to moral determination. National interests, realists maintain, necessarily displace morality, making foreign affairs essentially amoral. George Kennan, for example, writes that interests such as security, political independence and national wellbeing have no moral quality, as they are essential to the survival of states: 'They are the unavoidable necessities of national existence and therefore are subject to classification neither as "good" or "bad".'[4]

Realists resist applying morality to war. This resistance is part of their general moral scepticism that is applied to international relations overall:

The reason for the resistance is twofold. In the first place, it springs from the conviction that the reality in question is morally intractable, the dynamics of international relations and war being seen to confound most, if not all, attempts to apply an alien, moral structure to them. Secondly, and more urgently, it arises from the fear that the very attempt to impose a moral solution has tragic consequences. Not only does the attempt fail, it fails dangerously.[5]

Realism can be seen as a reaction against the tendency to apply moral norms to the international domain with little regard for the many constraints of the realities and complexities of power, or the intricate 'mechanisms whereby international order of an inferior but nonetheless real kind is sustained'.[6] The realist approach recognises both the possibilities and the constraints of power within an anarchic and bellicose world, as states strive to balance forces in order to bring some semblance of order, and since the balance of power involves forces in constant flux, maintaining it requires continual adjustment to changing circumstances free of moral constraints.

A pure form of the theory outright rejects the subjection of politics to ethics, instead affirming the radical autonomy of politics. Morgenthau, for example, defends 'the autonomy of the political sphere against its subversion by other modes of thought'.[7] 'The political realist', he argues, '[though] not unaware of the existence and relevance of standards of thought other than political ones ... cannot but subordinate these other standards to those of politics'.[8] Schlesinger attacks those who 'regard foreign policy as a branch of ethics'.[9] 'Realists', Carr suggests, 'hold that relations between states are governed solely by power and that morality plays no part in them'.[10] Well, not quite. The part that morality can play is instrumental. As Machiavelli argued in the sixteenth century, in *The Prince*, moral rhetoric is a powerful weapon in the statesman's armoury, and the ability to convey the appearance of virtue is an indispensable part of his art. The language of politics, particularly in time of war, is commonly a moral language and realists accept that even in power politics there is a place for morality.

Even so, aside from this instrumental role of morality in the hands of the statesman, international relations are thought to be morally indeterminable. If they cannot be morally ascertained during peacetime, they certainly cannot be morally ascertained in wartime, as war is international relations *in extremis*. In fact, when it comes to recourse to war – *jus ad bellum* – realism argues that morality is a poor guide. The decision to go to war should not be dictated by ideals and moral sentiment, but by considerations of power and interest. It would then follow that next to power and interest, considerations of justice and human rights would be equally poor as guides to action. As for the conduct of war – *jus in bello* – it is limitless: it can and should be fought by all available means: 'War is an act of violence pushed to its utmost bounds', Clausewitz wrote.[11] Trying to impose moral limits on a reality that is morally intractable is actually dangerous, because it will most likely lead to an increase rather than a reduction in human suffering, by, for example, prolonging war. 'In such dangerous things as War', wrote Clausewitz, 'the errors which proceed from a spirit of benevolence are the worst.'[12] In order to avoid such errors, any limitations on war must be non-moral; pragmatic, rational and strategic, rather than moral, considerations may guide conduct in war.

For realists war is not an end in itself, but a means to something else, the political ends that it is made to serve. As long as it retains that instrumental character it will remain limited by those political ends. Just as it was considered to be in a state's interests to go to war, so it can be judged to be in its interests to cease hostilities and to negotiate a peace. A war fought in accordance with realist principles, they argue, is much more likely to remain limited, than one fought on overt moral grounds. A 'moral' war can easily become a war fought without compromise, which can only end when the enemy's annihilation is achieved. The more moral or ideological a war becomes, the more it approaches the state of total war.

Since realism sees little chance for states to form successful alliances, as other states cannot ever be fully trusted, war is inevitable. Flexible rather than firm principles then are preferable. Realists view their flexibility as a virtue.

> It's a complex, rough-and-tumble world, they'd say, and countries would be foolish to take any option off the table for the sake of principle. After all, unscrupulous opponents, knowing such principles, may well use such scruples against you: for example by locating a militarily valuable target within a civilian population area, knowing that you won't attack it and it'll thereby be preserved. Or, worse, using child soldiers against you, knowing that you'd be extremely reluctant to strike at them. Everything ought to at least be considered.[13]

When a state is at war, it is only the safety of its own people that matters, or at least that takes priority over that of foreigners. Duties to distant strangers are at most voluntary and subject to pragmatic, national self-interest. Moral duties stop at the national border.

Clausewitz wrote about the changing nature of war in modern times. In the aftermath of the French Revolution, he argued, war underwent a transformation, coming much closer to its absolute form. No longer the affair of monarchs and their armies; with the birth of nationalism and the modern nation in the nineteenth century, war became an enterprise that engaged entire nations. The means employed in such a war were limitless, and its goal was nothing less than total victory, in keeping with the national myths and mission. Such an enthusiastic war would be immune to all attempts at moral limitation, and nation states would do whatever they could, in the midst of war, to try to win. Any peace achieved would be achieved through victory and through order.

There is another kind of war realism, Coates argues; a grassroots variety, not based on the ideas of theoreticians, the designs of strategists, or the plans of policymakers, but rooted in the experience of combatants:

> If realism applies at strategic or command level, then it applies with equal if not greater force at the level of combat. It would be foolish as well as hypocritical to expect soldiers to fight morally when the strategic and tactical planning that determines where and how they will fight is indifferent to moral considerations. The primary responsibility for the manner in which a war is conducted must lie with the political and military leadership that lays down the policy, strategy and tactics of war. However, even when the deliberations of the leaders have been formed in the light of moral imperatives and with moral intent, the exigencies of combat itself are often seen to be such as to vitiate the moral conduct of war.[14]

Part of military training necessarily involves preparing soldiers for combat in ways that undermine any moral capacity or inclination to fight justly or compassionately. The soldiers' emotional, mental and physical states are all geared towards self-preservation and enemy elimination. The effective soldier is dispossessed of his or her civilian and pacific responses, and turned into a killing machine, having been taught how to kill, but also having been given the desire to kill. Behaviours and attitudes that in peacetime would be regarded as unacceptable or even criminal, in war become the desirable norm. As Field Marshal Montgomery, senior British Army officer who served in the First World War, the Irish War of Independence and the Second World War, put it: 'The troops must be brought to a state of wild enthusiasm before the operation begins ... They must enter the fight with the light of battle in their eyes and definitely wanting to kill the enemy.'[15] If military training necessarily involves the nurturing of a homicidal constitution in the soldier, as a requirement of combat, a morally conducted war is neither desirable nor possible.

In addition, the practice of military discipline, as well as the existence and acceptance of an authoritarian and hierarchical command structure, mean that military training is also designed to elicit immediate and unquestioning obedience, while suppressing the critical reflection that moral judgement and conduct entail. At the same time, the nature of combat, which demands affinity and loyalty among comrades in arms, works against the moral conduct of war. As soldiers are thrown into shared danger and suffering, they must develop solidarity, which further narrows their moral horizons and deepens their moral sympathies. Their sense of moral obligation is, if not exclusively, then first and foremost towards each other. The stronger the obligation to the group is felt, the weaker the sense of obligation to those outside it. The killing of a comrade in arms may even unleash 'a blind hatred and a bloodlust that makes a mockery of the laws of war'.[16] So while a sense of morality or moral obligation exists in warfare, it is partisan and based on moral particularism, recognising only the moral claims of one's fellow combatants and excluding or ignoring those of adversaries.

War then cannot and must not be fought in a morally prescriptive or morally reflective way, according to this approach. Moral imperatives, moral considerations and moral reflections all place combatants at risk, jeopardising their safety and efficiency as fighters, and seriously affecting their mental and psychological state. Of course, soldiers are still human beings and as such they do not stop being moral agents, even during combat; yet war would become unbearable if they were expected to be morally self-critical during every battle or bombing campaign. The horrors and pressures of war, coupled with its moral ambiguity, mean that in order to survive it, protect one's comrades and defeat the enemy (through killing), moral doubt needs to be suppressed, or at least suspended. To remain mentally and psychologically stable (and militarily efficient), soldiers need to accept war's own logic.

The morally neutral, value-free realist understanding of war and international relations challenges the existence or application of ethics in war. Amoral realism suppresses the moral context of politics and of war. But another form of realism, while still resisting the moral determination of politics, is heedful of moral considerations. This type of realism recognises and affirms the moral dimension of politics, as well as the potential conflict between political objectives and morality, by suggesting that the achievement of political objectives necessitates the use of *immoral* means. In a crisis such as war, immoral acts are committed 'realistically'; a ruler may indeed have a duty to act immorally, though regrettably and with much moral anguish. This approach has been called 'Christian Realism' and it is associated with Reinhold Niebuhr, an American theologian and ethicist. This theological movement has its origins in the early 1930s, when D. C. Macintosh and Walter Marshall Horton wrote about religious realism or realistic theology, which influenced Niebuhr's call for a church that would produce religious or Christian realists.[17] Niebuhr's Christian and Protestant realism sees the political order as naturally resistant to morality, and the structure of power in which it consists as intrinsically flawed: it is impossible to act within the realm of politics without incurring sin. Though not a Christian realist, Morgenthau also argues that 'there is no escape from the evil of power' and that 'to know with despair that the political act is inevitably evil, and to act nonetheless, is moral courage'.[18] Both the politician and the soldier are faced with hard choices that require that they 'stoically immolate their personal morality on the altar of the public good'.[19] Michael Walzer, though a just war theorist, shares a similar view: 'Here is the moral politician: it is by his dirty hands that we know him. If he were a moral man a nothing else, his hands would not be dirty; if he were a politician and nothing else, he would pretend that they were clean.'[20] In the end, the conflict and irreconcilability of values that are involved in politics are inescapable. Moral principles are sacrificed at the altar of state sovereignty and national interests. In particular, as it does not distinguish between 'good' and 'bad' states, but rather between 'powerful' and 'weak', realism favours and legitimises the actions of powerful states. Orend writes:

If states ought to do what's in their best interests, and if the most powerful states have the greatest capability to do what they want, then the result is a legitimation of the fact that powerful states are able to exert such influence over the interstate system in general and smaller states in particular.[21]

This bias in favour of the powerful, Orend argues, is at odds with some of our other values, such as democracy and the self-determination of political communities. Even worse, it deprives smaller and weaker states of their most powerful claim for global reform: moral criticism. The danger of this bias is most evident in the case of great powers. Great powers, according to Mearsheimer, do not just want to be the strongest of all the great powers; they want to be the hegemon. For this reason, they are relentless in their pursuit of power, always looking for opportunities to alter the distribution of world power in their favour. 'They will seize these opportunities if they have the necessary capability. Simply put, great powers are primed for offense.'[22] If that is the case, all that morality would do, if imposed, would be to limit the ends that power seeks, and the means employed to achieve them.[23]

Wouldn't such limitations serve justice though? Especially during war? When it comes to justice, since realists believe in the primacy of self-interest over moral principle, and favour power and dominance in international relations, they may even regard considerations of justice as inappropriate, dangerous foundations on which to base foreign policies. Often appeals to justice justify or mask policies motivated by material interests. Influence can be achieved through *the appearance of justice*. When policy appears to be constrained by accepted ethical principles and to be supportive of them, it provides an aura of legitimacy that helps to reconcile weak actors to their subordinate status, as they become convinced that those to whom they must submit are acting in their interests or in those of the wider community.[24] Of course, influence can also be compelled by force (and realism deems this rational, if the capability is there), but influence obtained this way is expensive to maintain, tenuous in effect, and usually short-lived. A demonstrable commitment to justice can help powerful actors translate power into influence more effectively and long term. Justice is important for realists in another instrumental way: it provides the foundation on which powerful actors can construct interests:

In this respect, a commitment to justice is a powerful source of self-restraint, and restraint is necessary in direct proportion to one's power. Weak states must generally behave cautiously because of external constraints. Powerful states are not similarly restricted, and the past successes that made them powerful breed hubris, encourage their leaders to make inflated estimates of their ability to control events and seduce them into investing their assets and reputation in risky ventures ... Self-restraint that prompts behaviour in accord with the acknowledged principles of justice both earns and sustains the *hegemonia* that makes efficient influence possible.[25]

As for war, German historian Heinrich von Treitschke reminds us that 'without war, there would be no state', for all states are born of war; wars have been and will be fruitful.[26] And the state's most important task is to protect its citizens by arms.

Militarism

'Militarism is a condition in which war, or the preparation of the means of war, are major social concerns, commanding a significant proportion of resources and enjoying a substantial degree of legitimacy', writes MacKenzie.[27] Militarism is 'the propensity to use military power, or the threat of it, for political settlements'.[28]

During the Cold War a great deal of scholarly work was produced in the Western world on arms races, military expenditure, arms sales, and the militaristic attitudes, structures and practices that are shaped by warfare.[29] Extensive work was also undertaken on the concept of militarism itself. Discussion of militarism in the Cold War era had focused on the Soviet-American superpower rivalry, and on what Thompson labelled 'exterminism'.[30] With the end of the Cold War global military spending declined, and such research largely disappeared, briefly, only to be revitalised, renewed and readapted to the twenty-first century US-UK War on Terror, more recently in the context of the 'global surveillance war', where there was a shift in targets from traditional state threats and their militaries to much more amorphous terrorist networks and their affiliates.

The end of the Cold War was thought, by writers like Fukuyama, to have signalled the triumph of liberal democracy, the victory of liberalism and those peace-loving states that practise it. But did it bring peace? Liberal states actually have an unusually high propensity towards war with illiberal ones.[31] The leading liberal states of the nineteenth and twentieth centuries, the United Kingdom and the United States, fought more wars during these periods than any others.[32] The current leading liberal democratic state, the United States, accounts for over 40 per cent of global military spending[33] and has military personnel in more than 150 states.[34] Historically, liberal 'civil society militarism' visited genocide on large swathes of the non-European world.[35]

In the past 100 years militarism has been conceptualised in a number of ways. Militarism was understood as an *ideology* – one that glorified war, military institutions and the prevalence of martial values in society.[36] It put the emphasis on military considerations, ideals and values in the life of states. This type of definition became popular in relation to Germany around the time of the First World War, when militarism assumed 'the importance of a basic cultural value'.[37] Many late twentieth-century writers, from Mann[38] to Enloe,[39] defined militarism as involving a set of values and beliefs. A second understanding is behavioural, conceptualising militarism as the propensity to utilise force to resolve conflict.[40] Eide and Thee define militarism as 'the inclination to rely on military means of coercion for the handling of conflicts'.[41] Kinsella defines it as a disposition or proclivity to employ military over non-military means of conflict resolution. In addition, he writes, 'militarism in the contemporary era is already taking form as a greater willingness of major powers to intervene forcefully in the domestic affairs of other states'.[42] A third conception equates it with military *build-ups*, especially with quantitative increases in weapons production and imports, military personal and military expenditure.[43] *Sociological* understandings of militarism transcend the civil-military divide, by understanding militarism as embedded within society. Shaw refers to the penetration of social relations by military relations and the relationship of war preparation and society. In his work the concept of 'military' is used to describe 'all social relations, institutions and values relating to war and war preparation'.[44] Militarism, in turn, is understood as the tendency or extent to which these military relations influence social relations as a whole. The core idea is the 'carrying' of military forms into the civilian sphere.

'How is ethics implicated in the production of militarism?' James Eastwood asks in *Ethics as a Weapon of War*. Rather than provide reasons and ways to limit military violence, ethics has instead become part of an arrangement that makes violence easier for the military to commit. He sees the role of ethical activity in contemporary warfare as expanding and changing, and military ethics as staking profound claims on political legitimacy. In this environment, he argues, 'soldiering is presented as the acme of moral commitment':

> The turn to ethics has become useful for practitioners of military violence and corrosive of the critical and political capacity to engage with and resist violence … this is made possible by the strength and sophistication of *militarism*, an ideological force which is capable of mobilising ethical effort in the pursuit of military violence. Ethics can offer powerful support to militarism and, under certain circumstances, ethics can function as a weapon of war.[45]

Ethics is deployed in the service of militarism to create a moral army, claiming moral superiority over enemies that have been defined as evil, genocidal or terrorist. The 'moral army's' violence then is sanctioned, justified and portrayed primarily as defensive.

Militarism describes a particular interaction between military activity and social relations.[46] Militarism comes into being when using military force acquires legitimation, when it is perceived as a positive value and high principle that is right and desirable and is routinised and institutionalised on society.[47] It is also important to note that militarism, as Alfred Vagts observed, is not simply an enthusiastic love of war. So, militarism can be the state of being in thrall to war, of being influenced by, dependent on and caught up in military ways. War can become part of a wider set of social practices, structures and phenomena – for example, popular culture, or gender relations, which tie people's everyday activities, understandings and identities to military activity and the pursuit of war. Militarism can be understood as an ideological force, a set of discursive practices embedded in everyday life[48] and that serve to make war seem desirable and normal.[49] Fighting wars and preparing for war seem both normal and morally desirable, with ethics being integral to military preparations and fighting. The use of ethics is crucial to the ideological legitimation of military violence.

Armies use ethics to generate a widespread and prevalent ideology of ethical soldiering that instils a belief in a moral army with strong moral values. This is particularly important and useful in making the military attractive to recruits and in producing motivated soldiers: ethical soldiers with a strong sense of moral purpose and militarist identities. In the twenty-first century War on Terror, ethics have proven useful in the pursuit of counter insurgency or counter terrorist warfare.

For militarists, war is a matter of first preference, rather than of last resort. War is seen as capable of delivering essential goods to the state or nation that are not deliverable by peaceful means. According to Ceadel, underlying it is a form of moral particularism that excludes universal values and ruthlessly subordinates the good of humanity to the good of a particular race, state or nation.[50] Militarism involves excessive reliance on military power. A state of society that 'ranks military institutions and ways above the prevailing attitudes of civilian life and carries the military mentality into the civilian sphere'.[51] Militarism involves the glorification of military power and war.

Coates identifies a type of militarism that he calls 'crusading universalism', a form of imperialism. He argues that it is actually moral particularism masquerading as moral universalism:

> In the end 'crusading' too stands for the victory of one particular moral community over all the others, and the peace that it professes is dependent for its realization upon the forcible elimination of difference. The truth of the matter is that the crusader's 'altruism' ... is not the enemy but the ally of militarism. It is precisely the 'altruistic' pursuit of warfare that generates militarism and that leads to the systematic undermining of every limit placed upon war.[52]

The lust for war is a characteristic of militarism, especially the crusading type. Unlike the realist, who chooses war on pragmatic grounds, or the just war theorist, whose reluctant acceptance of the moral permissibility of war – on certain occasions and under certain conditions – stops short of moral enthusiasm, the militarist is a proponent of war. Militarism with its moral certitude creates an unobstructed path to war.

Militarism sees war as a positive good and as essential for human development. This is reflected in *Germany and the Next War* by

General Friedrich von Bernhardi, writing in 1912, 'The inevitableness, the idealism and the blessing of war, as an indispensable and stimulating law of development.'[53] There is a prestige attached to soldiering, an emphasis on the maintenance of a powerful army, and the constant reminder of threats to national survival. True values are martial values. Max Scheler, in *Der Genius des Krieges und der Deutsche Krieg* wrote:

> Mankind would have destroyed itself had not the dignity of war sanctified force and focused the attention of great peoples on common aims ... It is war which created existing societies out of potential ones and led directly to scientific and other developments being achieved ... The weapon preceded the tool ... It is war which justifies a noble people in their striving for increasing expansion and recognition.[54]

Anthony Ludovici, a prolific writer on the far-right fringe of British inter-war politics, accepted that 'a superior race has the incontestable right to spread itself at the cost of inferior races'.[55] Ludovici was primarily a crusader. War ethics are tied closely to a superior moral culture that must prevail. The true collective is the superior nation or state, which must, through war, strive to enlarge its sphere of influence. Imperialism is a form of crusading, as the superior state spreads its correct values in the colony. The imperialism in which liberal democracies freely indulged was (and is) itself militarist, since in respect of 'primitive' peoples it preached racial supremacy and the right of conquest.[56] In the twenty-first century that has become the right of indefinite strikes or targeted assassinations in states such as Pakistan, Afghanistan, Iraq, Syria, Yemen, Somalia and Libya. Crusading, an altruistic approach that aspires to promote a better world, resorts to coercion in the interests of order or justice. 'A positive crusade is motivated by a desire to impose new values or create a new system, whereas a negative one is designed to root out and destroy an existing evil.'[57] Ethics are at the core of such endeavours of moral arrogance to eradicate 'dangerous' regimes, while simultaneously imposing one's own values to colonial peoples that need to be civilised.

History abounds with examples of crusading militarism. This type of approach to war can be seen in the Christian crusades, in the US War on Terror, and in Islamic fundamentalism – where war becomes 'holy':

Islamic fundamentalism posits a state of war between the 'house or sphere of Islam' (dar al-Islam) and the 'house or sphere of war' (dar al-Harb), a state which is terminable only by the universal political hegemony of Islam. Legitimate dominion has one true foundation – Islam – and the recognition of non-Islamic states on anything other than a temporary and pragmatic basis is impermissible. This fundamentalist approach yields a starkly adversarial and conflictual view of the present state of humanity.[58]

Force is necessary for Islam to triumph, and it is enthusiastically proclaimed. As for war, it is not a necessary evil, but an obligation or duty.

This view of war is encountered in fundamental Islam's jihad and in the Christian world. As the militant Muslims of the Muslim Brotherhood or ISIS identify the good of humanity with the universal triumph of the Islamic caliphate, so militant Christians often identify it with the worldwide supremacy of a Christendom, politically and morally conceived. This can be seen in the American understanding of war as 'apocalyptic' and the assumption that the United States has 'a divine mission to save the world'.[59] For those who view war in such an exulted and morally uncompromising way, war has no contingent or instrumental character. As Aron has observed, 'the use of force with a view to such grandiose ends tends to become an end in itself'.[60] War is given a redemptive and transformative power, whereby a better humanity is created. In this way, war acquires a moral value far exceeding the instrumental value given to it by realists.

In such wars, 'ruthlessness is reinforced by the ethic of hardness that is common to all forms of militarism and that supplants the ethic of compassion that is so essential to the moral conduct of war'.[61] According to Benda, the extolling of harshness is rife among ideologues.[62] In this war ethic, compassion – ordinarily a virtue – is a sign of moral weakness rather than of strength.

The militarist understands war as total war, with unlimited conduct, an understanding that extends to the manner in which war ends. The militarist's war is fought not to redress specific injuries or to realise specific objectives; rather, it is a war that seeks an uncompromising and complete victory that involves the annihilation of an adversary and, ideally, the annihilation of the adversary's way of life. In such a war there is nothing an adversary can do, other than complete surrender, to terminate it. Militarism demands unconditional surrender or total submission. To settle for less would be seen as moral betrayal of the (moral) cause of war. In the 'holy war' of the Christian and Islamic traditions, victory or death is the necessary outcome, not mutual respect, negotiation or a bilateral settlement. Moreover, death incurred in the course of such a war has often been portrayed and praised (even sought) as a form of martyrdom.

Major states fight mainly against smaller states or armed groups – asymmetric war. Two types of actor have mobilised the new conditions to create new ways of war-making: Western states, especially the United States and the United Kingdom, and global Islamist terrorists:

> In what we may call 'the new Western way of war', the US, UK and (through international alliances and coalitions) other Western states have re-invented war-making, after the crisis of Vietnam, by packaging it as an inherently limited, temporarily and spatially bounded ('quick fix') affair.[63]

In these asymmetric wars, major states systematically transfer physical risk to local allies and, despite protestations of concern, to non-combatants. The new militarism deliberately avoids, above all, any deep social costs for Western societies: 'military policies are based strictly on professional force protection, designed to insulate Western economies, societies and polities from anything more than superficial effects'.[64] With careful media management – to either minimise or maximise the effects of mass killing – Western militarism focuses on societies/audiences who are mobilised through a patriotic/nationalist emphasis on force-protection, combined with global ideas of democracy-promotion and humanitarianism. Consequently, wars that are motivated by gain lead to perpetual conflict. 'The development of new constellations of interests is geared not to the ending of war but to its theoretically endless continuation.'[65] In the end, despite claims to ethicality, those that benefit from war are powerful states and military industries.

In 'War becomes academic' James Der Derian, interviewed by Stavrianakis and Selby, comments on the role of technology and ethics in the new 'virtuous' wars:

> The revolution in military affairs was never just about technology. It was also about culture and ethics. Technology enabled a conceit, helped

make a universalist virtue out of a particular self-interest. Technology made plausible the belief that through use of remote, precision munitions, you could fight a virtuous war, achieve political ends through military means, avoid the fog in the friction of war, achieve all of the things that Clausewitz warned us against. A revolution in military affairs would resolve political and ethical issues: war would once again become virtuous. 'Virtue' and 'virtual' shared a similar etymology, both conveyed the same meaning up until the Middle Ages, of producing divine effects at a distance, just as Jesus Christ did. But now we have a new *Deus ex machina*, network-centric warfare, producing the omnipotent effects of an alien power – Washington, DC, 6,000 miles away – perfectly targeting and taking out bad guys with Predator drones and Hellfire missiles. These are much more vengeful agents than Jesus Christ ever was. But it is all based on simulations and public dissimulations.[66]

The new virtuous war is a high-tech war fought from a distance, in both spatial and ethical senses. In the new virtuous/virtual/humanitarian war the declared aim is not to conquer, occupy, exploit or seize, but to fix, protect, save, secure and, while doing so, win hearts and minds. Geopolitics might still be operating beneath the surface, but increasingly wars – Western wars in particular – must be presented as having an ethical purpose that justifies an armed intervention. Modern war, through the militarist lens of virtue, is transformed into ideological-moral war (war that aims to vindicate a moral worldview), just like the religious wars that dominated an earlier age, generating militarism.

Just War

In contrast to realists, who regard war as essentially amoral and pragmatic, and militarists, who see war as having moral value, being sanctioned and guided by ethics, just war theorists insist 'on the moral determination of war where that is possible, and on the moral renunciation of war where it is not'.[67] Unlike militarists, proponents of just war affirm the moral primacy of peace over war, while simultaneously resisting the blanket condemnation of war and all things military, as they accept the potential moral instrumentality of war and the advantages (moral, instrumental and strategic) of an imperfect peace. The pure realist rejects the application of moral criteria to war as misplaced and even dangerous, regarding attempts to subject war to moral limitations as moralistic, misguided and ultimately risky. The just war theorist, on the other hand, insists on the moral determination of war, both in terms of recourse to war (*jus ad bellum*) and in terms of the subsequent conduct of war (*jus in bello*). Although for some realists moral considerations are applied to the question of recourse, once war has broken out moral considerations are not applied to the conduct of war, because the way that a war is fought is decided on military and political criteria, in relation to capability and strategy, and needs to be free from moral restraint. The just war tradition, however, upholds both a morality of ends and a morality of means. So, 'where the realist sees a continuum of politics and war, the just war theorist sees a radical disruption; and where the realist recognizes only pragmatic necessity, the just war theorist contemplates a moral tragedy'.[68] Realists such as Clausewitz recognised that there is a link between the ends that a war is meant to serve and the manner in which the war is conducted: the bigger or more ambitious the object of war is, the more unlimited is its conduct. The just war can never be unlimited or total, even when it is defensive. 'It is a crime to commit aggression, but aggressive war is a rule-governed activity. It is right to resist aggression, but the resistance is subject to moral (and legal) restraint. The dualism of *jus ad bellum* and *jus in bello* is at the heart of all that is most problematic in the moral reality of war.'[69]

As realists warn against moralism for its failure to acknowledge the realities of power and war, so just war theorists are critical of moralism, which they consider no better than amoralism. Like realism, just war theory understands the risks involved in subjecting war to morality, in the ways that militarists do. However, the just war tradition insists that moral criteria and constraints in war do not constitute moralism. The just war theorist resists the moral scepticism of the realist but accepts that the ethics of war must be informed by a realist perspective. What the just war tradition upholds is precisely a form of moral realism. Morality cannot be excluded from any aspect of international relations, including the extreme case of war, which is ultimately a human activity and as such informed by moral values.

The just war tradition is the predominant moral language through which we address the rights and wrongs of the use of force in international society. It provides us with a set of concepts and principles for understanding and responding to the moral and legal questions raised by war. As such, it is central to the theory and the practice of international relations. Its influence is evident in the legal codes that govern how modern militaries perform their duties, and it has featured prominently in the rhetoric surrounding the War on Terror and the military actions in Iraq, Afghanistan and Libya.[70] Just war adopts a universalist perspective in its analysis and moral understanding of security and war, within the context of an international community that both embraces and transcends all states, recognising *mutual* rights and obligations where realists argue that states have duties only to themselves.

The universalist/international perspective of just war theory can be seen in current international law, especially the laws of armed conflict. Just war theory and the laws of armed conflict apply abstract universal rules of conduct to warfare and hold the same core value convictions that war is sometimes permissible, and there's a difference between permissible and impermissible means of fighting it. While just war theory is much older, drawing its values and understandings from religious writings, ethical values, political debates and military experiences, the laws of armed conflict were constructed in the modern era by national governments, turning to just war theory for guidance.[71]

The laws of armed conflict stipulate that countries may resort to armed force only if these criteria of *jus as bellum* are met: just cause, proportionality, last resort and declaration of war by proper authority.

Just cause

'Just cause' is the most important rule that sets the tone for everything else that follows. When it comes to a just cause for going to war, international law (United Nations (UN) Charter, articles 39–53) recognises three general principles:

- All countries have the inherent, 'natural' right to go to war in self-defence.
- All countries have the inherent, 'natural' right of other-defence – to go to war as an act of aid to any country victimised by aggression.
- Any other use of force is not in the eyes of international law an inherent, 'natural' right of states. Any country wishing to

engage in more controversial forms of force (such as a preemptive strike) needs the approval of the UN Security Council. Failing to secure prior authorisation renders any such use of force illegal, itself an act of aggression.

Proportionality

Since war is costly, bloody and unpredictable, only a very few problems would be so bad that war would be a proportionate response. The laws of armed conflict command that the problem in question be serious enough that war is a proper, fitting reply. So what problem is so severe that a declaration of war is justified? International law's answer is *aggression*:

> When confronted with an aggressive invader ... which is intent on conquering and essentially enslaving other nations, it's deemed reasonable to stand up to such a dark threat to life and liberty and to resist it and beat it back, with force if need be. Just as dangerous criminals must be resisted and prevented from getting away with their crimes – lest chaos be invited – countries are entitled to stand up to aggressors, and to defeat them.[72]

Last resort

All attempts to resolve the issue in question (including aggression) with the tools of foreign policy need to be exhausted, before resorting to war. Those tools include diplomatic negotiations and economic incentivising. Orend advises that the meaning of this rule is relative to the full context of the situation, such as the gravity of the threat, the nature and actions of the aggressor, and the preferences and capabilities of the victim and any of its allies.

Proper authority

Only the branch of government that has the war power, has the authority to publicly order the use of force and warfare. Only the governments of sovereign states have legitimate authority to wage war. In the twenty-first century a problem arises around authority regarding the use of drones within the War on Terror and cyber conflict, both new forms of force that involve secrecy or evade public scrutiny. For this reason, they are susceptible to abuse by those with the war power, which is what this rule of proper authority is meant to prevent.

Further considerations of just war theory

Two further considerations fall within the scope of just war theory, although they are not included in the laws of armed conflict: right intention and probability of success.

Right intention
The right intention (created by St Augustine, 354–430), whereby a ruler (i.e., the proper authority) ordering war must do so only out of love for his people, who are defenceless and in need of his protection, and only with the greatest reluctance. Any personal or subjective intentions, any motives such as greed (for power or resources) and hatred (for the enemy) do not constitute right intention. Love, the desire to protect, to provide defence from aggression: these are the only morally right or permissible motives behind a decision to go to war.

Probability of success
It is wrong to go to war if you know that there is a high probability that you will lose. Engaging in something that will bring a great deal of death and destruction, to combatants and civilians, and that will end in defeat, is both irrational and insensitive to human suffering.

Conduct During War (*jus in bello*)

How are belligerents supposed to fight? Realists say that once war has begun, in theory, everything is permitted; in practice, it may be wise to restrain oneself. It depends on the nature of the enemy, the level of threat, their intention and capability, one's own military capability, and what is more likely to contribute to one's objective in the circumstances. Both just war theory and the laws of armed conflict, in contrast, impose firm permissions and prohibitions regarding what belligerents may do during armed conflict. The overall function of *jus in bello* is two-fold: (1) to uphold a kind of ideal regarding a fair and decent way to fight; and (2) to put clean limits on the amount and kind of force being deployed, in particular so that total war (unlimited, indiscriminate escalation in violence) does not happen. Without limits or restrictions, war crimes such as ethnic cleansing, genocide, or the use of weapons of mass destruction may occur. Hundreds of laws of armed conflict serve to realise six principles of just war fighting:

1. Discrimination and non-combatant immunity;
2. Benevolent quarantine for prisoners of war;
3. Proportionality;
4. No use of prohibited weapons;
5. No use of means '*mala in se*' ('evil in itself' – acts inherently immoral, such as murder or rape); and
6. No reprisals.

Discrimination means the need for fighters to distinguish, or discriminate, between legitimate and illegitimate targets, and to take aim with armed force only at legitimate targets. Illegitimate targets include residential areas, schools, hospitals, farms, sites of religious worship, cultural institutions and non-military industrial sites. Non-combatants – 'protected persons' – are considered immune from intentional attack. The intentional killing of civilians is considered the worst war crime.

International humanitarian law is synonymous with *jus in bello*:

> It seeks to minimize suffering in armed conflicts, notably by protecting and assisting all victims of armed conflict to the greatest extent possible. IHL [international humanitarian law] applies to the belligerent parties irrespective of the reasons for the conflict or the justness of the causes for which they are fighting. If it were otherwise, implementing the law would be impossible, since every party would claim to be a victim of aggression. Moreover, IHL is intended to protect victims of armed conflicts regardless of party affiliation. That is why *jus in bello* must remain independent of *jus ad bellum*.[73]

Just war theory and international humanitarian law ultimately emphasise the international community's moral and legal responsibility to prevent violations like war crimes. Moreover, international humanitarian law is intended to protect war victims and their fundamental rights, no matter to which party they belong.

2 • Killing from Afar: The Terror of the Air

By the start of the Second World War, technologies had transformed the ways in which the air was lived, conceived and perceived, as a space of risk, opportunity and power. As a result of the revolutionary changes to the nature of warfare, the air became territorialised, the privilege of the powerful. Wartime bombardment became everyday domination: a reminder of the constant presence of state power. The air became a volume to occupy and a means to exercise terror. It also led to the tragic consequence of the pilots not being quite aware of whom they had killed on the ground, as killing became a distant affair. While twentieth-century technology enabled more efficient mass killing, inflicting maximum damage to the enemy, while danger-proofing one's own military personnel, the ethics of warfare became blurred.

Air power is important in war because control of the air is essential for armies and navies to perform effectively and to win battles. The First World War saw a marked quickening of aviation technology, and by its conclusion in 1918 the fundamental roles and missions of air services had been established. The success of aircraft in the conflict also led to deeper thinking about military aviation, and from this stemmed a range of controversies and debates over the fundamental purpose, legality and morality of air attacks. 'The nature of air power is a contested area. It swiftly became the subject of intense debate as to its precise utility and its political, ethical and technological dimensions.'[1]

The main attributes of air power are reach and rapidity. Aircraft can engage the enemy at depth, through strikes against rear-echelon forces (an element of a military headquarters located far from the front and concerned with administrative and supply duties) or by attacking war industries or communications infrastructure, supporting the belief that strategic bombardment could win wars without ground forces.[2] The flexibility of aircraft combined with the rapidity with which they can reach the battle area, enabled commanders to concentrate air power and maximise the psychological and material impact. In the absence of significant air defence, aircraft can attack targets almost at will. In modern warfare success cannot be guaranteed without control of the air, and the likelihood of loss and failure is dramatically increased.

In August 1945, US forces dropped two atomic bombs on Japan's Hiroshima and Nagasaki, killing hundreds of thousands of civilians, providing an exceptionally bloody end to an extremely brutal war, but also showing that the only means of delivering nuclear weapons was from aircraft. This prompted both sides in the Cold War that followed to invest in their air defence assets, while 'the US Air Force remained wedded to the notion that air power could win wars alone'.[3] The concept of 'Rapid Dominance' was developed, in which air power would play a key part in overwhelming the enemy with great rapidity, and the employment of air power was seen as the most effective way to coerce opponents, forcing them to change their behaviour.

This chapter looks at air warfare during the Second World War, focusing on aerial bombardments by the United States, the United Kingdom and Germany, as well as the toxic air of death camps. It provides a critical examination of remote killing: the advantages, consequences and morality of danger-proofing at the physical and psychological levels. We take a close look at what the soldiers did not see.

The Bomber Wars

The bombing of London and other British cities by the German Luftwaffe began in 1940. As part of Operation Sea Lion, General Alfred Jodl, Chief of Operations Staff (sentenced to death and executed at Nuremberg in 1946), wanted to increase pressure on Britain to agree to a negotiated peace. The first priority was to gain air supremacy, which meant defeating the Royal Air Force (RAF). Intense air attacks on UK cities could affect supplies and civilian morale, and bring quicker capitulation. Terror raids and the depletion of Britain's food stocks would, it was hoped, pound the civilian population into submission and force the British government to surrender. The full force of German bombing was felt by London's docks, the City, Whitehall, the East End and south London, where the factories were located. When British civilian casualties started to mount, Berlin would claim that they were 'collateral damage' and reprisals for RAF bombing of German civilians. The German bombing campaign of the United Kingdom, known as 'the Blitz', lasted from September 1940 until the German invasion of Russia in May 1941. The result of the bombing was extensive destruction: one raid alone, on 29 December 1940, almost obliterated the City of London. By the end of that period, there were 30,000 British dead and 50,000 injured. The first day of the Blitz, 7 September 1940, is remembered as Black Saturday. Beginning on Black Saturday, London was attacked on fifty-seven straight nights.

London was not alone in receiving the visitations of the Luftwaffe from 1940 to 1941. Notable among the German air attacks were those on Coventry (bombed forty-one times during the war), the deadliest of which was on 14–15 November 1940. The raid on Coventry lasted ten hours and more than 500 bombers were involved. The damage was extensive: its fourteenth-century cathedral was destroyed, its tram system smashed, its gas and water supplies put out of action, 4,500 houses demolished and 600 of its citizens killed. The Germans claimed to have been targeting the many military-related engineering works in the city, but the high altitudes that they had to maintain in order to avoid anti-aircraft fire made accurate bombing impossible. Bombs fell everywhere on the city indiscriminately. German propaganda after the attack said that it was a retaliation for RAF bombing of Munich. In early 1941 the Germans launched another wave of air raids on Plymouth, Portsmouth, Bristol, Newcastle upon Tyne, Hull, Swansea, Belfast and Clydeside.

The Blitz on Liverpool, 1–7 May 1941, was the most concentrated series of air attacks on any British city area outside London during the Second World War. The impact on civilians was devastating. During a bombing raid on 4 May 1941, Margaret Johnson took her children, Peter (aged fifteen), John (aged seven) and Margaret (six months) to safety in a shelter near their home in Wykham Street, Kirkdale. As the shelter was bombed, Margaret desperately tried to protect her baby daughter by kneeling over her, to shield her from the falling debris. Tragically, all three children died: the baby suffocated, John was killed instantly, while older brother Peter was severely injured and died six days later, on 10 May 1941. Margaret was injured.[4]

The Blitz was such a horrific bombing campaign, that it made the idea of severe retaliation against German cities more acceptable, even welcome. On 24 July 1943, a fleet of 791 RAF bombers took off from their bases, on a mission to unleash an assault on the German city of Hamburg and its citizens. The assault on the city, named Operation Gomorrah and mounted by the RAF bomber Command with the aim of wiping Hamburg off the map, lasted a week and a half. It was reported that Hamburg's fire fighters were overwhelmed by the torrents of incendiaries that fell on to the city, so many and in such concentration that they initiated a terrifying phenomenon: a firestorm. In his book *Among the Dead Cities*, Grayling writes that the bombing started fires in different streets that joined together, forming:

> vast pyres of flame that grew rapidly hotter and eventually roared upwards to a height of 7,000 feet ... It was the first ever firestorm created by bombing and it caused terrible destruction and loss of life. Its greatest intensity lasted for three hours, snatching up roofs, trees and burning human bodies and sending them whirling into the air. The fires leaped up behind collapsing facades of buildings, roared through the streets, and rolled across squares and open areas ... The glass windows of tramcars melted, bags of sugar boiled, people trying to flee the oven-like heat of air raid shelters sank, petrified into grotesque gestures, into the boiling asphalt of the streets.[5]

Some bodies were found so shrivelled by the heat that adult corpses had shrunk to the size of infants. At least 45,000 corpses lay among the burnt and smoking ruins, with thousands more injured and traumatised. Sebald, in his book *On the Natural History of Destruction*, provides a shocking detail in an account given by someone who saw refugees from Hamburg trying to board a train in Bavaria. In the struggle, a woman drops a suitcase that 'falls on the platform, bursts open and spills its contents. Toys, a manicure case, singed underwear. And last of all, the roasted corpse of a child, shrunk like a mummy, which its half-deranged mother has been carrying about with her.'[6] A woman who hurried to Hamburg the day after the firestorm to find her parents, gave the following account:

> Women and children were so charred as to be unrecognisable; those that had died through lack of oxygen were half charred and recognisable. Their brains tumbled from their burst temples and their insides

from the soft parts under the ribs. How terribly must these people have died. The smallest children lay like fried eels on the pavement.[7]

When the bombers left Hamburg on the night of 24 July, they could not see the roasted or shrivelled corpses of adults and children, they could not smell the burning flesh or witness the trauma inflicted on survivors; all they could see was a city burning, and they could see it from 120 miles away. Two nights later they were back, and they were back yet again a further three nights after that. The greatest percentage of bombs dropped in the entire war, and the greatest destruction of German and Japanese cities, occurred in the war's final months, when the war was already won.

In September 1940, British Prime Minister Winston Churchill, in a memo to the Cabinet, said 'our supreme effort must be to gain overwhelming mastery in the air. The fighters are our salvation, but the bombers alone can provide the means to victory.'[8] Between 1941 and 1945, the British bombed Hamburg, Dresden, Berlin, Munich, Lubeck, Augsburg, Cologne and Bremen, killing an estimated 600,000 civilians and injuring a further 800,000.[9]

The atomic bombings of Japan's Hiroshima and Nagasaki by the United States in August 1945 remain the biggest mass killings in history and the only use of nuclear weapons in armed conflict, but the air raids over Japan had started earlier. An air attack against Tokyo took place on the night of 9–10 March 1945. Tokyo received 1,667 tons of incendiary bombs on fifteen square miles of its most densely populated districts. The ferocious firestorm that was created killed 185,000 people. The total tonnage of bombs dropped by Allied planes in the Pacific war was 656,400. Of this, 160,800 tons, or 24 per cent, were dropped on the home islands of Japan. Total civilian casualties in Japan, as a result of nine months of air attacks, including those from the atomic bombs, were approximately 806,000. Of these, approximately 330,000 were fatalities, according to the United States Strategic Bombing Surveys, which included the Pacific Wars Summary Report, first published in July 1946. These casualties exceeded Japan's combat casualties, which the Japanese estimate as having totalled approximately 780,000 during the entire war. The principal cause of civilian death or injury was burns.

Some of those who entered the Hiroshima and Nagasaki after the bombings to provide assistance also died from radiation. While a fireball from a nuclear explosion takes only ten seconds to reach its maximum size, the effects can last for years and span across several generations. Just a few years after the bombings, survivors developed leukaemia in great numbers; they also started to suffer from thyroid, breast, lung and other cancers at higher-than-normal rates. Pregnant women that had been exposed to the bombings had higher rates of miscarriage and deaths among their babies and young children; the children were more likely to have intellectual disabilities, impaired growth and an increased risk of developing cancer. For all survivors of the atomic bombings, 'cancers related to radiation exposure still continue to increase throughout their lifespan, even to this day'.[10]

Air bombing kills more civilians and destroys more cultural heritage than ground war does.

For and Against the Bombings

The International Peace Conference at The Hague in 1899 involved the first-ever effort to restrict aerial bombing, in order to see civilians unharmed as far as possible in war. Debates about the dangers anticipated from aerial bombing had begun. The conference was convened on the initiative of the Czar of Russia, Nicholas II, 'with the object of seeking the most effective means of ensuring to all peoples the benefits of a real and lasting peace, and, above all, of limiting the progressive development of existing armaments'.[11] Twenty-six governments were represented in the conference, including Germany, Italy, Japan, the United Kingdom and the United States, and they failed to reach agreement on the limitation or reduction of armaments. A second conference in 1907 led to constitute authoritative statements of the results achieved, which were signed by the delegates but not ratified by the participating states. As a result, they had no binding force.

The fullest and most thoughtful attempt to provide rules for air warfare was made at a conference held at The Hague between December 1922 and February 1923. The then five major powers of Britain, France, the United States, Italy and Japan took part. The articles drawn up by the conference were never signed by the participating governments, so no rules of air warfare came into existence. However, the draft rules arrived at by the participants show how clearly the dangers of air power were foreseen. As Grayling points out, what happened in conflicts after the powers considered these rules, and especially what happened in the Second World War, must be measured against the principles that they embody. Articles XXII and XXIV are particularly relevant:

> Article XXII
> Aerial bombardment for the purpose of terrorising the civilian population, of destroying or damaging private property not of a military character, or of injuring non-combatants, is prohibited.
>
> Article XXIV
> Aerial bombardment is legitimate only when directed at a military objective.

When the Geneva Disarmament Conference (the Conference for the Reduction and Limitation of Armament) began in February 1932, a year before Adolf Hitler was appointed as Chancellor, most of the attending powers were in agreement that air strikes on civilians were in violation of fundamental principles; but the conference 'stalled on the political realities of the time'.[12] Germany claimed the right to build up its armed forces and in 1933 announced its withdrawal from both the Disarmament Conference and the League of Nations. Meanwhile, in November 1932, Stanley Baldwin (three times British prime minister between 1923 and 1937) made a famous speech in which he warned that 'the bomber will always get through'. His speech is titled 'A Fear for the Future', and he speaks of 'the appalling speed which the air has brought into modern warfare':

> In the next war you will find that any town within reach of an aerodrome can be bombed within the first five minutes of war to an extent inconceivable in the last War, and the question is, whose morale will be shattered quickest by that preliminary bombing? I think it is well also for the man in the street to realize that there is no power on earth that can protect him from being bombed, whatever people may tell him. The bomber will always get through, and it is very easy to understand that if you realize the area of space. Take any large town you like on this island or on the Continent within reach of an aerodrome. For the defence of that town and its suburbs you have to split up the air into sectors for defence. Calculate that the bombing aeroplanes will be at

least 20,000ft. high in the air, and perhaps higher, and it is a matter of mathematical calculation that you will have sectors of from 10 to hundreds of cubic miles. Imagine 100 cubic miles covered with cloud and fog, and you can calculate how many aeroplanes you would have to throw into that to have much chance of catching odd aeroplanes as they fly through it. It cannot be done, and there is no expert in Europe who will say that it can. The only defence is in offence, which means that you have got to kill more women and children more quickly than the enemy if you want to save yourselves. I mention that so that people may realize what is waiting for them when the next war comes.[13]

This part though, where he speaks of the 'terror of the air', was the most powerful and prophetic:

We have to remember that aerial warfare is still in its infancy, and its potentialities are incalculable and inconceivable. How have the nations tried to deal with this terror of the air? I confess that the more I have studied this question the more depressed I have been at the perfectly futile attempts that have been made to deal with this problem. The amount of time that has been wasted at Geneva in discussing questions such as the reduction of the size of aeroplanes, the prohibition of bombardment of the civil population, the prohibition of bombing, has really reduced me to despair. What would be the only object of reducing the size of aeroplanes? So long as we are working at this form of warfare every scientific man in the country will immediately turn to making a high-explosive bomb about the size of a walnut and as powerful as a bomb of big dimensions, and our last fate may be just as bad as the first.[14]

A fair warning, just before the horrors unleashed by a fast-approaching world war. Yet the prohibition of the bombardment of civilians was considered impracticable, and the bombings that followed necessary. In the House of Commons, on 14 September 1939, Prime Minister Chamberlain declared, 'His Majesty's Government will never resort to the deliberate attack on women and children and other civilians for the purpose of mere terrorism'.[15] However, there was no ban on such attacks; indeed, the RAF objected to any such ban, as bombers were linked to the very survival of the RAF. As long as it was not for the purpose of mere terrorism, but linked to some important war objective, even the bombing of civilians could go ahead.

While not illegal, was the deliberate military attack on civilian populations a moral crime? There are moral questions arising about area bombing (otherwise called 'carpet bombing', 'saturation bombing', 'obliteration bombing' and 'mass bombing'). Are there circumstances in which killing civilians in wartime is not a moral crime? Are there ever circumstances that would justify them?

An argument can be made that bombing civilians protected soldiers' lives. Bomber Command turned to area bombing because it found precision bombing too difficult and too dangerous for the pilots. Such an argument would be compatible with the realist approach, which places great value on a state's military, the state's most important task being the protection of its territory, its interests and its citizens by arms. Protecting the primary means of security must be the ultimate aim, making soldiers the greatest commodity in wartime. Moral considerations and reflections would only place soldiers at even greater risk than they already are. It could even be argued that civilian losses are not as important as military losses, so sacrificing the civilian – from a safe distance for the soldier – seems to be a correct war tactic. The decision to go to war is always dictated by considerations of power and interest, in any case, so we cannot expect armies to fight morally, but to use all available means to win that war, and the first and necessary condition for this is for them stay alive. Any type of warfare that protects those who must be kept alive has to be rational, strategic and in the state's best interest.

Civilians (ours and theirs) always die in war; premature or violent death is inevitable. The survival of the state is the priority and that can be ensured only through a victory that the state's security forces will provide. Moreover, a state's selfish and relative view of international relations, the view that states have duties only to themselves, greatly diminishes in value the lives of civilians in other states. Which makes foreign civilians, in times of war, even less valuable than domestic ones for the state. For the Allies, the lives of German and Japanese men, women and children were, at best, a regrettable but necessary casualty of war; at worst, of little to no value. Just as, for the Germans, the thousands of British lives lost in the Blitz would have been a necessary casualty of a war they aimed to win.

The ultimate aim is to win the war. In the Second World War it was vital for both sides to use bombing to win, knowing that winning a war by city bombing involved the mass killing of civilians and a morally intractable victory. The saying 'to make a wasteland and call it peace', with 'victory' substituted for or added to 'peace', applies here.[16] It is particularly poignant in the case of the atomic bombings, which forced the Japanese surrender. A world war ended, victory was secured and peace was restored – but at what cost?

Robin Neillands, in *The Bomber War: The Allied Air Offensive Against Germany*, cites Clausewitz and Macaulay in support of the idea that once war has broken out the only aim is to win it 'at any cost, and especially if that cost can be met by the enemy'.[17] According to Clausewitz, the will of the enemy is a legitimate target: 'the destruction of his capacity to resist, the killing of his courage rather than his men' is what warfare is about. For Macaulay, writing in 1831, 'The essence of war is violence; moderation in war is imbecility'.[18] For Neillands, the ethics of war can be endlessly debated by moral philosophers, but wars are not fought by moral philosophers. Wars are fought by ordinary people, by military commanders and by those put in positions of power and responsibility to their country. All of whom, during a war – especially a world war – would find moral questions largely academic. Their aims are to stay alive and to win.

Has morality any place in war at all? War involves an endless list of atrocities, a lot of which were committed during the Second World War. Could just war theory provide a moral evaluation of city bombing? The just war tradition does not condemn war, but insists on the moral determination of war, in terms of recognising and addressing rights and wrongs in the use of force by militaries. When it comes to recourse to war (*jus ad bellum*), any action by Germany or by Japan during the Second World War would be unjust and immoral, as they fought a series of aggressive wars, invading and occupying territories in Europe, Asia and Africa, killing and victimising millions of people. 'A war waged for such reasons as self-interest, *Lebensraum*, other people's oil fields, or pure aggrandisement, is not justified.'[19] The Allies, on the other hand, fought a defensive war, which according to this theory is just, as countries have the right to go to war in self-defence, as well as in aid or to the rescue of people subjected to aggression.

When it comes to the conduct of war (*jus in bello*), it gets a little less clear. While the Allies fought a just war of defence, guided by proper authority, and as a last resort, arguably with the right intention and having a good probability of success – all of which satisfy just war ethics – was the deliberate bombing of hundreds of thousands of civilians necessary or proportionate? For realists war may be unlimited, but just war proponents recognise that we have mutual rights and obligations, which means that war cannot be without limit, but must respect and honour the duties to humanity. The bombings were so many, so prolonged and so high in casualties, it cannot be argued that any one of them was necessary or effective, or they would not have continued for years. Rather, the air raids over so many cities, for so long, causing such tremendous loss of life and destruction of all that makes life possible, look like reprisals. As such, they were immoral, a moral wrong violating civilised standards of treatment of human beings. That a war is just does not necessarily make right every act committed during the fighting of it. Acts of injustice can be perpetrated even in the course of a just war, and if the injustices committed are themselves very great, their commission threatens the overall justice of the war.

Maybe there are some things that should never be attacked or harmed, even if doing so helps to bring about victory. The concept of innocence is important, argues Grayling. The Latin word *nocens* means 'engaged in harmful activity'; the prefix *in* means 'not'. So, *in-nocens* means 'not engaged in harmful activity': 'This idea imposes an obligation to distinguish between combatants and non-combatants, and they must not intend to cause harm to the latter either as a means to their ends, or as an end in itself, in their conduct of the war.'[20] There must be morality in war, according to just war theory, or war becomes aggrandised barbarism. While difficult, the effort to apply morality in war must not be abandoned. The air attacks that targeted civilians were moral atrocities, aimed at causing maximum hurt, disruption and terror to the innocent.

The importance of ethics is emphasised in militarism too, but in this case a moral conviction in the role, superiority and aims of soldiering is what makes even extreme violence easier for the military to commit. Ethics is deployed to create a moral army whose violence is sanctioned. Military ways are ethical ways, while military violence is not necessary or strategic, but desirable and ideologically legitimate. The ethical soldier with a strong moral purpose and strong moral values cannot possibly be committing immoral acts in war. That could even be extended to conducting deadly airstrikes over urban areas.

Militarism does not endorse a universal morality, but a relative one. The glorification of military power goes hand in hand with the glorification of the particular moral community (such as a nation) that is superior to all others. Crusading that generates militarism undermines every limit placed on war and justifies a noble people in their struggle. Any consideration of the welfare of the 'other' is considered moral weakness.

Militarism was a powerful force in twentieth-century Europe. While British militarism was more subdued than its German counterpart, military power was and still is considered essential for maintaining Britain's imperial and trade interests. In addition, soldiers are heroes and their actions during war must be nothing less than heroic, even as they bomb civilians. Rather than condemn them for committing war crimes, the nation must instead continuously acknowledge the sacrifice and selfless service that soldiers make. The unrestrained militarism of Nazi Germany was not unique to the Germans: militarism can be seen in other historical examples of nationalist uprisings, imperialisms and general ideological renderings, 'as parts of respective philosophies of history'.[21]

The militarisation of diplomacy, foreign policy and even domestic politics lay at the heart of Chancellor Bismarck's practice of *Realpolitik*, in a game of balance of power with England, France and Russia. He is remembered for his militarisation of the new German Empire's state and society, with his war mobilisation plans, economic coordination, press propaganda and pursuit of overseas imperialism.[22]

> A myth of untrammelled military greatness as a substitute for diplomatic excellence that the Second Kaiser and, ultimately, Adolf Hitler would tie themselves to leading directly to the destruction of the Second and Third Reich's. In other words, the lesson learned by Bismarck's foreign affairs was not *realpolitik*, not a diplomatist balance of power centered foreign policy, but *machtpolitik*, a military-centered preponderance of power perspective.[23]

Bismarck institutionalised the myth of German invincibility and militaristic crusader spirit, thus increasing the role of the 'myth' in the social consciousness of the German people. It was tapped into by Wilhelm II at the start of the Great War and you can see it waxed philosophically in the newspapers and even academic treatises of the time. The idea is captured in Hitler's speeches and writings as notions of 'blood soil', 'racial greatness' and 'fatherland volk'.[24] The theme of order promoted by the Nazis was employed by conservative, nationalist, militarist, monarchist and other reactionary groups in speeches, propaganda and demonstrations. In Germany, the Nazis emerged at the top of the race to remake the world, within a militarist tradition of parades, demonstrations, symbols and speeches, all of which were used by any number of groups throughout the country. It is worth noting that they were also used in other countries, including the United Kingdom and the United States.

Hitler's fascination with the military and his belief his own military genius were part of the general inclination across German society towards the virtues of military training and activity in daily living in the Germany of the late nineteenth century and early twentieth century.[25] One of Hitler's favourite quotes captures this view:

> War was a good in itself because states, like people, were driven to dominate each other and warfare was the process by which national domination was achieved ... That war should ever be banished from the world is not only absurd, but profoundly immoral ... Permanent peace would be a crime, and societies that wanted peace were obviously decaying.[26]

Unbridled militarism, where the military becomes the end rather than the means of politics, against which Clausewitz had argued, leads to a not only untroubled march to war, but also to an unquestioning acceptance of war as a just endeavour. This ideology justified the mass killings of the bomber wars, where the air advantage was capitalised on by all sides, creating a risky space of terror above and a bloody destruction on the ground that the bomber pilots, hailed as heroes, did not see. As the air became another territory to occupy by the powerful, a space from which they killed and terrorised the vulnerable, another atrocity was taking place: the gassing of millions of civilians in extermination camps.

The Toxic Air of the Death Camps

In 1942, when he was twenty-one years old, Oskar Gröning, German SS Unterscharführer, was posted to Auschwitz. According to him, between 80 per cent and 90 per cent of those on the first transport that he witnessed in September 1942 were selected to be murdered at once. When it was over, it was 'just like a fairground':

> There was a load of rubbish, and next to this rubbish were ill people, unable to walk, perhaps a child that had lost its mother, or perhaps during searching the train somebody had hidden. And these people were simply killed with a shot through the head. The kind of way in which these people were treated brought me doubt and outrage. A child was simply pulled by the leg and thrown on a lorry ... then when it cried like a sick chicken, they chucked it against the edge of the lorry. I couldn't understand that an SS man would take a child and throw its head against the side of a lorry ... or kill them by shooting them and then throw them on a lorry like a sack of wheat.[27]

Gröning, filled with outrage at the brutality, complained to his superior officer, 'If it is necessary to exterminate the Jews, then at least it should be done within a certain framework.' His superior officer replied that the excesses he saw that night were an exception, and that he agreed that members of the SS should not participate in such 'sadistic' events.[28]

In March 1943 the first gas chamber became operational at Auschwitz, to murder Jews and others persecuted by the Nazis. In all, four large gas chambers and crematoria, which included incinerators for burning the bodies, became operational in March, April and June 1943. The Nazis calculated that the crematoria could burn up to 4,416 corpses per day, but prisoners who worked there thought that the figure may have been almost double that.[29] Indeed, much like airstrikes, death camps provided the SS with certain advantages: they could kill thousands of people in one day, and they could do so while sparing the German soldiers a lot of the horror involved in exterminating terrified, defenceless families. The latter was achieved by using non-Germans in the actual removal and disposal of the bodies, after they had been gassed out of sight.

Three categories of people worked at the death camps. The first consisted of Jews. Christian Wirth, German SS officer at Belzec, realised that employing the Jewish prisoners (*Sonderkommandos*) in the killing process would spare his own men psychological suffering or emotional trauma, but would also mean that fewer Germans would be needed to run the camp. So, hundreds of fit, healthy and able-bodied Jews were selected from the arriving transports and put to work burying the bodies, cleaning the gas chambers and sorting the enormous quantity of clothes and other belongings that piled high in the camp. The second category of workers were Ukrainian guards. Famed for their brutality, many of them had previously fought for the Red Army. And then there were the Germans, the third category:

> But so smoothly had Wirth delegated the mechanics of running his killing machine to other nationalities that only 20 or so German SS needed to be involved at Belzec in the process of murder. By March 1942, with the arrival of the first transport at Belzec, Wirth had realised Himmler's dream. He had built a killing factory capable of exterminating hundreds of thousands which could be run by a handful of Germans, all of whom were now relatively protected from the psychological damage that had afflicted the firing squads in the East.[30]

Efficiency, maximum impact, emotional danger-proofing and delegation were all factors in the decision to use this killing method in the systematic and mechanised extermination of millions. Those who witnessed the horrors of the gas chambers were primarily the *Sonderkommandos*, groups of Jewish prisoners forced to perform a variety of horrific duties in the gas chambers and crematoria, including the removal and disposal of the corpses. The *Sonderkommandos* were forced to perform tasks of unimaginable horror and brutality, before facing death themselves:

> The Jews of this work brigade fulfilled many roles in the gas chamber/ crematoria complex, beginning with the arrival of victims. *Sonderkommando* Jews were present in the undressing area, instructing arriving victims to undress and how to arrange their clothing, which would later be confiscated by the SS. After the gassing was carried out by camp personnel, men of the *Sonderkommando* entered the gas chamber, untangled bodies, and cleaned the by then ventilated room. An elevator then raised the bodies to the crematorium. There, another group would shave the hair of the victims and search the bodies for hidden valuables, including gold teeth, which were removed and handed over to the SS. Finally, the *Sonderkommando* Jews would carry out the burning of bodies in the ovens and the disposal of the ashes.[31]

To destroy the evidence of mass murder, they were also forced to exhume bodies from burial pits and burn them. The use of the *Sonderkommando* thus relieved SS personnel from the most disturbing and psychologically destructive tasks.

Dario Gabbai and Morris Venezia, cousins from the Greek city Thessaloniki, were two such prisoners transported to Auschwitz, caught in the Nazi's *Sonderkommando* recruitment drive:

> The gas chambers of crematoria 2 and 3 were below ground, so the delivery of the poisonous Zyklon B, once the chamber was crammed with people and the door secured, was relatively straightforward. Standing outside on the gas chamber roof, members of the SS would take off hatches that gave them access to special wire columns in the gas chamber below. They would then place canisters of Zyklon B inside the columns and lower them, sealing the hatch again once the gas had reached the bottom. From the other side of the locked door, Dario Gabbai and Morris Venezia heard children and their mothers crying and scratching the walls. Morris remembers how, when the gas chamber was crammed with around a thousand people, he heard voices calling out. 'God! God!' 'Like a voice from the catacombs, I still hear this kind of voice in my head.' After the noise ceased, powerful fans were turned on to remove the gas, and then it was time for Morris, Dario and the other Sonderkommando to go to work. 'When they opened the door,' says Dario, 'I see these people that half an hour before were going [into the gas chamber], I see them all standing up, some black and blue from the gas. No place where to go. Dead. If I close my eyes, the only thing I see is standing up, women with children in their hands.'[32]

When they re-entered the gas chamber to pry the dead bodies apart with hooked poles and to clean up the blood and excrement from the walls and floor, they were supervised by as few as two SS men. This limited to a minimum the number of Germans who might be subjected to the kind of psychological damage that members of the killing squads in the East had suffered.

The air was weaponised once again, this time with a poisonous gas. Zyklon B, the brand name for hydrogen cyanide, was used originally as a disinfectant and insecticide, but it proved to be an efficient and deadly murder weapon during the Holocaust. By 1941, the Nazis had already decided to kill Jews on a mass scale. They looked for the fastest way to accomplish their goal. 'After the Nazi invasion of the Soviet Union, *Einsatzgruppen* (mobile killing squads) followed behind the army in order to round up and murder large numbers of Jews by mass shootings, such as at Babi Yar', writes Rosenberg. 'It wasn't long before the Nazis decided that shooting was costly, slow, and took too high a mental toll on the killers.'[33] Gas vans were tried as part of the Euthanasia Programme, the systematic murder of patients with disabilities, and at the Chelmno Death Camp. Carbon monoxide exhaust fumes from trucks killed the people crammed into the enclosed back area. These killings took an hour to complete. Höss, commandant of Auschwitz, and Eichmann, a German officer in charge of exterminating Jews and others, searched for a faster way to kill and decided to try Zyklon B. On 3 September 1941, 600 Soviet prisoners of war and 250 Polish prisoners no longer able to work were forced into the basement of Block 11 at Auschwitz I, known as the 'death block', and Zyklon B was released inside, killing them within minutes. Days later, 900 Soviet prisoners of war were sent to a large room at Crematorium I at Auschwitz for 'disinfection'. Zyklon B pellets were released from a hole in the ceiling – the prisoners died quickly. In five to twenty minutes, depending on the weather, all the people inside suffocated. After everyone had died, the poisonous air was pumped out, which took roughly fifteen minutes.

Zyklon B proved to be an effective, efficient, quick and cheap way to kill large numbers of people. Moreover, this method of killing allowed for the extermination of millions with little impact on military personnel, the vast majority of whom never saw the black and blue bodies, or burnt the corpses of children and babies that had suffocated in their mothers' arms, from which they had to be prised with hooked poles.

Killing Remotely

Killing those you cannot see is easier. Technology made it possible to hunt prey and fight enemies from great distances, but also in ways that psychologically distance the perpetrators of violence from their victims, providing a 'moral buffer', thus reducing the overall impact of killing.[34] A sense of distance, both physically and emotionally, is a significant element in the development of a moral buffer. This moral buffer allows people to ethically distance themselves from their actions and diminish a sense of accountability and moral responsibility. In a 2017 study, California State University researchers Rutchick, McManus, Barth, Youmans, Ainsworth and Goukassian argue that killing remotely is psychologically easier than killing face to face, which could promote more killing behaviour and could incur less severe emotional consequences.[35] In an experiment that they conducted, participants completed an ostensible ladybird (or 'ladybug') killing task. The participants who were in the same room as the insects killed fewer of them than those who killed remotely. Remoteness also influenced self-reported emotional consequences of killing.

Rutchick's experimental study was one of very few on the psychological effects of remoteness on a person's decision to take a life. The research used an insect-killing paradigm to provide experimental evidence of the effect of remoteness on killing and its emotional consequences. While his study involved the 'killing' of ladybirds, Rutchick said the results raise questions about how aware those making the decisions to kill remotely – such as military personnel who oversee air and drone strikes – are about the influences on those decisions to kill or not kill. 'The use of technology and distance has always been a part of killing – after all, someone realized long ago that it was easier to kill throwing a spear than up close – but this, a new medium of technology, is qualitatively different', Rutchick believed.[36] He wondered if the distance could have a psychological impact. The study measured whether people found it easier to kill when there was a distance between themselves and their victims. They created a remote-controlled machine that purported to killed ladybirds and more than 330 California State University, Northridge (CSUN) undergraduate students were recruited to test it. The students were told that the machine could be used to produce biological samples or dye at an industrial scale. After being shown how to use the remote control to operate a conveyor belt to carry the ladybirds to a box that contained a grinder that would kill them, the students were divided into three groups:

> those who operated the machine in the same room as the machine; those who were told the machine was in another location in California and operated it via Skype; and those who were told the machine was in Virginia and operated it via Skype. Each group could 'kill' as many ladybugs as they liked, but had to kill at least two of the insects to ensure they conducted a 'good' test of the machine.[37]

The findings of the study were that (a) participants killed more ladybirds when they were in different rooms from their targets, (b) participants who killed more, and were thus engaged in the killing task for longer, reported experiencing fewer negative emotions about what they did, (c) the location of the machine – whether in Virginia or California – made little difference to those operating the machine from a distance. When faced with moral dilemmas, people find decision-making harder, if the process is more intimate and personal. The study demonstrated the impact of psychological distance on cognition and moral judgement, and the ways they manifest in behaviour. Rutchick made clear that he did not intend to tell military leaders how to do their job, but the findings of this study, he hoped, would 'inspire a conversation – when people are considering a lethal decision – about what is impacting that decision'.[38]

People are less reluctant to end the life of another being when they are not physically present for the distasteful act. Although executing fellow humans is rather different from slaying an adorable insect like a ladybird, the study constitutes the first experimental evidence that remoteness increases killing behaviour. Knowing this reality, 'it is important that we grasp the psychological consequences of committing annihilation from afar'.[39]

Military psychology scholars have argued that intimate killing is psychologically difficult. Soldiers who experience direct combat are more likely to experience maladaptive mental-health outcomes[40] including post-traumatic stress and suicidality.[41] Training soldiers to fight and kill entails overcoming a natural aversion to life-taking,[42] however, killing from elsewhere entails fewer such inhibitions.

Decreasing intimacy between aggressors and targets can also increase aggression. In Milgram's studies of obedience, when remoteness between the teacher and learner was greater (no audio connection versus audio connection versus same room versus physical contact), the learner inflicted more harm. Milgram's obedience studies of the 1960s illustrate how the concept of remoteness from the consequences of one's actions can influence human behaviour in a negative, unethical fashion. The focus of Milgram's obedience research was to determine how a legitimate authority could influence the behaviour of subjects, and if the subjects would be obedient to requests, even if they could cause serious physical harm. The subjects thought that the purpose of the research was to study learning and memory, and that they would be the 'teachers'. As teachers, the subjects would administer increasing levels of electric shocks to a 'learner', who was actually a confederate participant, when this person made mistakes on a memory test. The trial most pertinent to this discussion of moral buffers is the experiment in which the teacher could or could not see the learner. When the learner was in sight of the teacher, 70 per cent of the subjects refused to administer the shocks, compared to the 35 per cent who resisted when the subject was remotely located.[43] Milgram argued that several factors could influence the tendency to resist shocking another human, when the human was in sight, one of which was the idea of empathetic cues. When someone is remotely administering painful stimuli to other humans, they are aware only in a conceptual sense that suffering could result from their actions, which creates emotional distance. A sense of moral superiority, perceived racial and ethnic differences, can contribute to emotional distance. Another factor that accounts for the distance/obedience effect is narrowing of the cognitive field for subjects: the 'out of sight, out of mind' phenomenon. In addition, the actual physical separation between an action and the resultant consequences is another factor that contributes to the ability to inflict greater suffering through remote conditions. All of these are factors in the formation of a moral buffer.

Neurological evidence shows that moral judgements involving personal closeness are processed differently from other judgements. Greene, Sommerville, Nystrom, Darley and Cohen examined a version of the 'trolley problem' in which participants decide between allowing a runaway trolley to kill five people and sacrificing one person to prevent the five deaths.[44] Greene *et al.* found that, when people considered personal dilemmas (e.g., pushing a man on the track to stop the trolley), reaction times were slower and different brain regions were activated, than when people considered fewer personal dilemmas (e.g., pulling a switch to divert the trolley).

The Morality of the Gas Chambers

Death came through the air for those who died in the gas chambers, in killings that, for the soldiers, were accomplished through physical and emotional distance: distance from the actions and from the consequences of those actions. The delegation of the disposal of the bodies completed the 'clean' executions, protecting the SS in every way from their abhorrent deeds, at the same time achieving maximum impact.

Just war theory would consider immoral the airstrikes that killed hundreds of thousands of people, both those conducted by the Germans, and those conducted by the Allies – which killed a lot more civilians, in Germany and in Japan. Realism could provide justifications for all airstrikes; if not moral, then strategic, in keeping with war tactics and with the desire of all states involved to win the war. Neither of them, however, would justify the poisoning of people in gas chambers. As Germany was the aggressor, just war proponents would argue that all its actions in the Second World War were unjust and immoral. From a realist perspective, the systematic killing of millions in death camps (Jews, Gypsies and those suffering from mental illness or physical disability), served no strategic interest in the war, or was ever in any way a contributing factor in Germany's war outcomes, or in restoring/sustaining political order within its society. There is no justification whatsoever, on any grounds, for those acts.

The war on the Jews, Gypsies and the disabled is not justified by either of these theories, but is it justified by militarism? Did the Nazi construction of an 'enemy' or threat to society create a narrative that justified the extermination of civilians as normal and moral? We can reach a possible justification through a type of militarism called racial militarism. Racism is 'the belief, practice, and policy of domination based on the specious concept of race'[45] and is often supported by state power. Meanwhile, militarism 'justifies military priorities and military influences in cultural, economic, and political affairs'[46] and includes an 'ideological legitimation of violence'.[47] 'This legitimized violence is, like racism, always for the purpose of domination. Moreover, given its monopoly over violence, the state (as is the case with racism) is the most effective agent of militarism.'[48] The term 'racial militarism' denotes the overlapping functions of domination and legitimation of violence. 'Given their similar functions, one can already see how morbidly useful racism and militarism are to each other, and how the state operates as a conduit for both.'[49] Jasmine Gani argues that race was present at the inception of European militarism and is still perpetuated through its continued performance *vis-à-vis* racialised communities. According to her, racial militarism operates as a theory of civilisational supremacy and as a practice/policy of chauvinism, exclusion and dehumanisation, for the purpose of enacting violence. More generally, we can argue that racial militarism considers normal and ethical the elimination of these 'inferior' beings that constitute a threat to civilised society. Through racial militarism the state dominates its barbarian rivals to restore order to the civilisational hierarchy. This theory and practice is associated with stigmas regarding intellectual, moral and physical inferiority.

War Ethics Post-Second World War

The laws of war and humanitarian declarations following the Second World War were framed expressly to address the atrocities that had been committed. While the crime that was most urgently addressed was that of genocide, the new laws, declarations and war ethics also sought to protect the rights of the individual, to outlaw military aggression and the destruction of cultural heritage. The Geneva Convention 1949 and its protocols specifically formulated the principles of protection for civilians in time of war, which were based on those of just war theory. It was decided that states must give unequivocal instructions to their armed forces not to mount direct attacks against civilians or civilian targets, indiscriminate attacks, or attacks which, though aimed at legitimate military objectives, have a disproportionate impact on civilians in the vicinity. Additionally, states must not use weapons that are inherently indiscriminate in their effects, but they must take all necessary measures to protect civilian populations from the effect of military operations. The Nuremberg trials demonstrated the moral and legal efforts to infuse humanitarian considerations into *jus in bello*. The high command in wars (military and political) – that is, those who made the decisions – were put under greater responsibility to show consideration and adopt a more encompassing view of war: not just of strategy and tactics, but also of the political, diplomatic and human implications of what is involved in the decision to go to war, and in the conduct of war.

Yet questions around the conflicts involved in setting ethics against politics persisted. Should morality take priority over political reality, or should the political (interests and power) take priority over what is right or good? In *Ethics and the Limits of Philosophy*, Bernard Williams writes:

> we naturally say that some things are morally important, others aesthetically important, and so on. But there must be a question at the end, in a particular case or more generally, whether one type of importance is more important than another type.[50]

While Williams grants that people must rely as far possible on something as fundamental as not being killed or used as a resource, he argues that the ideal that human existence can be ultimately just is illusory.[51] Williams argues that problems of political ethics need to be discussed within their peculiar field of application that constitutes their own ethical constituency.

Rather than just war principles guiding action, Gray argues that realism's ambivalent ethics are more appropriate:

> Realism is the only way of thinking about issues of tyranny and freedom, war and peace that can truly claim not to be based on faith and, despite its reputation for amorality, the only one that is ethically serious ... Realism requires a discipline of thought that may be too austere for a culture that prizes psychological comfort above anything else, and it is a reasonable question whether western liberal societies are capable of the moral effort that is involved in setting aside hopes of world-transformation.[52]

History, Gray maintains, is not a story of moral progress, but one of gains and losses:

> History is not an ascending spiral of human advance, or even an inch-by-inch crawl to a better world. It is an unending cycle in which changing knowledge interacts with unchanging human needs. Freedom is recurrently won and lost in an alternation that includes long periods of anarchy and tyranny, and there is no reason to suppose that this cycle will ever end.[53]

While it can be argued that international law after the Second World War gave priority to moral and humanitarian principles consistent with just war, the moral limitations inherent in the political ends of states led to a 'politics first' type of morality, in peace and in war. Power politics remained largely unrestrained from moral limitations and the Weinberger doctrine in the United States emphasised the full commitment of US troops to winning wars, when vital to the national interest. More alarmingly, where ethics remained strong was in militarism.

Significantly, despite the many trials for war crimes committed during the Second World War, there were no trials for the hundreds of thousands of civilians killed through air attacks, by any of the warring parties. This is important, because air warfare was allowed to continue, to develop and to remain the preferred mode of fighting. The air continued to be a space of risk, of danger and of opportunities, part of a political geography of distance. The subsequent development of unmanned aerial vehicles (UAVs) led to new ways for states to exercise power, control and terror.

3 • From the Bomber to the Drone

> I dropped my son at school in the morning, continued on to work and, within a couple of hours, killed two men. I went home later that day to be greeted by my son with a cheery, 'How was your day?'
>
> JAY, REAPER PILOT[1]

'This is the story of a machine', starts Neillands's book *The Bomber War*. A strategic and tactical weapon. A machine that could win wars by itself, 'the bomber dream' that could be made a reality in the skies over much of Europe. But the strategic bomber was a failure:

> In the first quarter-century of its existence the bomber lacked the navigational aids, the target-marking and bomb-aiming equipment, and the bomb-carrying capacity which would enable it to make a significant contribution to the outcome of any major struggle. As a weapon of war the strategic bomber was totally inadequate.[2]

The bomber was not the unstoppable weapon or ultimate deterrent it was thought to be, but a destroyer of cities and civilian lives, a weapon of terror, posing moral questions such as: were civilians a legitimate target in war? If, for example, it was legitimate to bomb arms factories and to kill the soldiers who used the arms, why was it unethical to bomb the homes of people who made those weapons? After all, wars in the twentieth century were fought by whole nations. General Curtis LeMay, speaking after the Second World War about the bomber offensive, said, 'To worry about the morality of what we were doing? Nuts! A soldier has to fight', adding:

> If you are cursed with any imagination at all, you have at least one horrid glimpse of a child in bed with a ton of masonry tumbling down on top of him, or a three-year-old girl crying 'Mutter, mutter', because she

has been burned. You have to turn away from that picture if you intend to retain your sanity and do the work your nation expects of you.[3]

Inspiring terror, causing mass destruction and mass killings, was what the strategic bomber was for. What it could not do was destroy precision targets and prevent wars. Also, bombers could not fully protect the aircrew, who, although greatly distanced from the killing, were still at risk. Everyone who flew was afraid, according to Howard Jackson, a bombardier with the USAAF 454th Bomb Group, Fifteenth Air Force:

> The terror starts on the night before the mission. This should not be confused with fear. Fear is when you have to ask a girl to dance who might say no, or when waiting in class to be asked a question you don't know how to answer. Terror is anxiety, dreams, rationalization of excuses not to fly, headaches, loose bowels, shaking and silence.[4]

A new weapon had to be developed, one that would render previous weapons and tactics obsolete, providing full moral, mental and physical buffering to the pilots, while enabling them to conduct precision strikes, or targeted assassinations, that would prevent total war and save hundreds of thousands of lives. In the new post-Second World War era of human rights and international law protecting the civilian, a new 'ethical weapon' came in the form of the drone. This weapon, it was thought, would be in perfect concordance with the principles of the just war theory (on which the Geneva Conventions and international humanitarian law were based), doing away with moral problems and dilemmas concerning *jus in bello*.

This chapter explores the history, development and morality of killer drones.

Drones and War: The History of Unmanned Aircraft

Drone prehistory goes back to 1849, when bomb-dropping balloons were used in the siege of Venice, after a revolt against Austrian rule. Those killer balloons were devised by Uchatius, Lieutenant in the Austrian army. It was not possible to bring siege artillery close to the city, so Uchatius rigged up twenty hot-air balloons to release small bombs by remote control via a copper wire. Austrian news reports suggested the bombs would turn Venice into rubble, but only one or two hit the city, the rest falling 'into the waters of the Venetian Lido or outside the city entirely'.[5] However, the first air raids in history by unmanned balloons carrying bombs, as they floated over Venice, led to the extreme terror of its inhabitants.[6]

In the United States, the first pilotless aircraft was developed during the First World War. In 1915, with the war raging in Europe, a board of inventors were helping prepare the United States for possible entry into the conflict. Among the board's members were scientist-inventors Peter Cooper Hewitt and Elmer Sperry, who began development of an 'aerial torpedo', together with Sperry's son Lawrence. By 1916 the father-son team had produced a workable 'automatic pilot' system[7] and Lawrence Sperry filed for a patent for an 'aerial torpedo' – essentially an unmanned aircraft. The aircraft would have a wind-driven generator to provide power for the gyro motors and servomotors that would move the control surfaces. The Naval Consulting Board approved the Sperry-Hewitt aerial torpedo project in April 1917, the same month the United States entered the First World War. By 1918, the aerial torpedo was able to fly along a pre-set route and dive on a target, delivering a thousand-pound bomb or releasing a torpedo. Meanwhile, the US Army was developing its own aerial torpedo under the leadership of Charles F. Kettering, who sought to produce a cheaper, simpler unmanned aircraft. He called on Orville Wright to help develop the airframe for his aerial torpedo. Flight tests of the 'Kettering Bug' began in September 1918, and after some successful flight tests, the army ordered 100 of the Kettering aerial torpedoes. However, interest in both projects waned over the next few years and the pilotless aircraft was not given much consideration until the late 1930s, with the US Navy developing the Naval Aircraft Factory/Interstate TDN-1 unmanned aircraft and both the US Navy and German Air Force developing air-launched guided bombs.

In December 1944 unidentified flying objects were observed on the US west coast. On investigation, these were found to be 30-foot-long paper balloons filled with hydrogen and containing a complex mechanical gondola. At first, they were thought to be weather balloons, but after reports of unexplained explosions, one was captured intact and found to be carrying incendiary bombs. This was the Japanese Fu-Go or 'windship weapon'. The balloons had flown all the way from Japan, using the jet stream, a narrow ribbon of fast-moving air at high altitudes. The Fu-Go was a drone, rather than an unguided missile, because of its ingenious method of altitude control:

> A clockwork mechanism controlled the release of a jet of small sandbags around the rim of the gondola. Whenever the balloon fell too low, it dropped another sandbag. If it rose too high, which might cause it to burst, a valve vented a small amount of hydrogen. This control system meant it maintained height and stayed in the jet stream for the three-day journey across the Pacific. Once it had expended the sandbags, the Fu-Go started dropping its load of incendiary bombs, one by one. The aim was to start forest fires in the heavily wooded regions of the Pacific Northwest. This would spread panic and divert resources from the war effort. The target was big enough that even this rough method of aiming had a chance of success. The simple balloons were made from mulberry paper and vegetable glue. Nine thousand were assembled by Japanese schoolgirls, many working in sports halls normally used for sumo bouts. Some of the workers were so hungry that they ate the glue they were working with. US analysts estimated the Fu-Go cost $200 each, at a time when a P-51 Mustang was $50,000. Of the 9,000 balloons launched by the Japanese, experts estimated that perhaps 900 reached North America.[8]

Though not very effective, the Japanese balloons did cause some casualties on the ground, to a party of Sunday-school picnickers. On Saturday 5 May 1945, as the war in Europe was ending and just months before the Japanese surrender, 'spinning shards of metal ripped into the tall pine trees, burrowing holes into bark and tearing needles from branches outside the tiny logging community of Bly, Oregon. The nerve-shattering echo of an exploding bomb rolled across the mountain landscape.'[9] The bomb killed six people, one of them Elsie Mitchell, the minister's pregnant wife. The rest were children: Jay Gifford, Eddie Engen, Dick Patzke, Sherman Shoemaker and Joan Patzke.

The earliest UAVs were known as remotely piloted vehicles (RPVs); they were small radio-controlled aircraft and fell into two categories: small, inexpensive and often expendable vehicles used for training; and, from the 1950s, larger and more sophisticated systems recovered by radio-controlled landing. They were fitted with reflectors to simulate the radar return of enemy aircraft and used as decoys to help bombers penetrate enemy defences. RPVs were also used for photographic and electronic reconnaissance. The AQM-34 Firebee was built in 1951 by the Ryan Aeronautical Company in the United States. First flown in 1962, the reconnaissance Firebee saw extensive service in the Vietnam War, North Korea and China. Drones also started to be used in new roles, such as acting as decoys in combat, launching missiles and dropping leaflets for psychological operations.

The genesis of the modern drone came in key technological leaps. First came the work of Abraham Karem, an Israeli aviator, who in the 1970s started to develop aircraft with glider-like properties. Karem, the 'drone father', is regarded as the founding father of UAV (drone) technology, having created the robotic plane that transformed modern warfare. 'I was not the guy who put missiles on the Predator', he said. 'I just wanted UAVs to perform to the same standards of safety, reliability and performance as manned aircraft.'[10] When Karem left Israel for the United States in 1977, the Pentagon had almost given up on robotic planes. At the time its most promising UAV, the Aquila, needed thirty people to launch it, it flew for just minutes at a time and crashed on average every twenty flight hours. Karem developed his initial drone during the Yom Kippur war for the Israeli air force, and in the United States he constructed the Albatross and later the more complicated Amber, which became the celebrated Predator.

Drones were lighter than manned aircraft and did not have to land when they got tired, they could loiter for long periods, while the pilots stayed on the ground, keeping an eye on the targets. The second advance was the use of transmitters to send the footage straight back to battlefield commanders. The signal to control them, and the returning video footage, are transmitted through satellite networks. In 2000, the United States took the final leap forward when the Air Force and the CIA became the first to successfully fit drones with missiles.[11]

UAVs were critical to the success of NATO's Operation Allied Force in 1999. The Army's Hunter UAV and the Navy's Pioneer drone were used throughout the seventy-eight-day air campaign in Kosovo, gathering intelligence and data on bomb damage. In the case of the Hunter, produced in the mid-1990s by Israel Aircraft

Industries and operated by the US Army's Task Force Hawk in Macedonia, target planners were able to watch live video feed of the Serbian soldiers in Kosovo. The US Navy deployed its own Israeli-designed UAV to support its missions. Navy officials credited the Pioneers with having provided NATO commanders valuable intelligence during operations, flying reconnaissance missions over Kosovo and Serbia. 'These rugged fiberglass aerial platforms are much cheaper than traditional piloted aircraft and each one lost on a reconnaissance patrol may represent a pilot's life that was saved by not having to undertake the same mission.'[12]

Control of the air was crucial in NATO's attack on Yugoslavia, to enforce acceptance of a solution to the Kosovo crisis. After suffering severe losses, the Yugoslav Air Force reduced flying to a bare minimum, resulting in a NATO success. An estimated 25,000 houses and apartment buildings were destroyed in the bombing, as well as 470 kilometres of roads and 600 kilometres of railway. According to Serbia's defence ministry, NATO forces killed 631 members of the Serbian armed forces. The number of casualties remains unclear, and Human Rights Watch puts the civilian death toll at around 500. NATO has never revealed the losses.[13] Eighteen years later, Serbian Prime Minister Aleksandar Vucic described NATO's military campaign as 'one of the most brutal aggressions in the history of warfare'. The airstrike campaign in Yugoslavia – then Serbia and Montenegro – was launched without a UN Security Council mandate. The best-known mass killing of the campaign was the hitting of a passenger train by two missiles, near Grdelica, Serbia, on 12 April 1999; the airstrikes killed at least twenty civilian passengers. In Montenegro, six civilians, including three children, were killed in an airstrike on a bridge in the village of Murino on 30 April 1999.

The key incidents that were covered by the Final Report, on page 5, were:

a. the attack on a civilian passenger train at the Grdelica Gorge – 12/4/99 – 10 or more civilians killed, 15 or more injured
b. the attack on the Djakovica Convoy – 14/4/99 – 70–75 civilians killed, 100 or more injured
c. the attack on Surdulica – 27/4/99 – 11 civilians killed, 100 or more injured
d. the attack on Cuprija – 8/4/99 – 1 civilian killed, 5 injured

e. the attack on the Cigota Medical Institute – 8/4/99 – 3 civilians killed
f. the attack on Hotels Baciste and Putnik – 13/4/99 – 1 civilian killed
g. the attacks on the Pancevo Petrochemical Complex and Fertilizer Company – 15/4/99 and 18/4/99 – no reported civilian casualties
h. the attack on the Nis Tobacco Factory – 18/4/99 – no reported civilian casualties
i. the attack on the Djakovica Refugee Camp – 21/4/99 – 5 civilians killed, 16–19 injured
j. the attack on a bus at Lužane – 1/5/99 39 civilians killed
k. the attack on a bus at Pec – 3/5/99 – 17 civilians killed, 44 injured
l. the attack at Korisa village – 13/5/99 – 48–87 civilians killed
m. the attack on the Belgrade TV and Radio Station – 23/4/99 – 16 civilians killed
n. the attack on the Chinese Embassy in Belgrade – 7/5/99 – 3 civilians killed, 15 injured
o. attack on Nis City Centre and Hospital – 7/5/99 – 13 civilians killed, 60 injured
p. attack on Istok Prison – 21/5/99 – at least 19 civilians killed
q. attack on Belgrade Hospital – 20/5/99 – 3 civilians killed, several injured
r. attack on Surdulica Sanatorium – 30/5/99 – 23 killed, many injured
s. attack on journalists convoy Prizren-Brezovica Road – 31/5/99 – 1 civilian killed – 3 injured
t. attack on Belgrade Heating Plant – 4/4/99 – 1 killed
u. attacks on Trade and Industry Targets.[14]

Criticism of the NATO bombing campaign has included the allegation that, as the resort to force was illegal, all NATO actions were illegal. Also, that NATO forces deliberately and unlawfully attacked civilian infrastructure targets, deliberately or recklessly attacked civilians, and deliberately or recklessly caused excessive civilian casualties in disregard of the rule of proportionality. 'Certain allegations went so far as to accuse NATO of crimes against humanity and genocide.'[15]

The most dramatic losses of civilian life from the NATO offensive in Kosovo came from attacks on fleeing or travelling refugees confused with military forces. These included repeated attacks on refugees over a 12-mile stretch of the Djakovica-Decane road in Kosovo, in which seventy-three civilian refugees died; attacks near Korisa, in which as many as eighty-seven civilian displaced persons and refugees died; and two incidents involving attacks on civilian buses, at Lužane and Savine Vode. Drones started life as spy planes and became assassination weapons.

Until the 1980s drones were machines used mainly for intelligence, surveillance and reconnaissance. They were just eyes, not weapons. Their transformation took place between Kosovo in 1999 and Afghanistan in 2001, as the new millennium began. No more than a few months before 11 September 2001, officers who had seen the Predator at work in Kosovo had the idea of equipping it with an antitank missile. Bill Yenne, in his history of the drone, writes:

> On February 16, 2001, during tests at Nellis Air Force Base, a Predator successfully fired a Hellfire AGM-114C into a target. The notion of turning the Predator into a predator had been realized. No one could imagine that, before the year was out, the Predator would be preying upon live targets in Afghanistan.[16]

In the aftermath of the 9/11, attacks, as the War on Terror was about to spread to Iraq, George W. Bush declared that the United States would embark on a new kind of warfare, 'a war that requires us to be on an international manhunt'.[17]

Drones come in different forms: terrestrial drones, marine drones, submarine drones, subterranean drones. Provided that there is no human crew aboard, any vehicle can be 'dronised'. A drone can be controlled either from a distance by human operators, or automatically by robotic means. Present-day drones combine the two modes of control. Armed flying drones are known as 'hunter-killers'; their history is that of an eye turned into a weapon. A good definition of drones is 'flying, high-resolution video cameras armed with missiles'.[18] The advantage of unmanned aerial systems is that they allow the projection of power (deploying military force) without projecting vulnerability (vulnerable bodies are out of reach), making them both effective and ethical, thus enabling 'good' wars.

In *Wired for War*, Singer argues that the 'good' wars often hide the seamy underside of war. 'The history of warfare is a history that often shames us and it should', he argues, because 'war is human destruction, horror and waste all wrapped together.'[19] The great historian Thucydides described war as a punishment springing from man's hubris, but war has also been described as the only true place where man's virtue, nobility and excellence could be tested. At the start of the fifth century BCE, the Ionian philosopher Heraclitus observed that 'war is the father of all things'. His statement revealed a belief that stood the test time:

> Like Heraclitus and after him Herodotus, we too see war as 'the ultimate expression and at the same time the ultimate theatre for the "display" of "great and astonishing achievements", a theatre where success and failure are visible to all and awesome in their consequences'. It was Herodotus' goal that these 'great and astonishing achievements' should not be forgotten, as he notes in the preface to his *Histories*. Yet while he recognised the heroism of men in battle, he was also an acute and critical observer of the horror of war and of its ironies.[20]

War reveals our faults, but also our strengths. From Homer to Shakespeare, to John Le Carre, virtues like bravery and self-sacrifice are taught through stories of war. War is also depicted as immoral, 'yet humanity has always found out-clauses to explain its necessity and celebration. The same religions that see violence as a sin also licenced wars of crusade and jihad.'[21] It is the same in politics: we urge war as a means to spread ideologies and 'good' systems such as 'Western values' and democracy, or defeat them, as in the case of fascism, communism and religious extremism. What is observable in war through the stories that we tell about it is also found in the narratives surrounding the wars we fight remotely, since science fiction became 'science reality'.

Creating mechanical beings to replace the work of humans is an old idea, going back to ancient Greece. Hephaestus, the god of metalwork in Greek mythology, had mechanical servants that he made out of bronze. One of them was Talos, a huge automaton that Hephaestus gave to the Cretans, to protect the island of Crete from pirates and invaders by throwing large boulders at approaching ships. He circled the island's shores three times a day. If a stranger made it ashore, Talos would heat up his metallic arms and give the intruder a deadly hug. Talos was the name for an Apple computer operating system, as well as the first computer-guided missiles on US Navy ships.

These myths inspired philosophers, as well as inventors. In 322 BCE the Greek philosopher Aristotle wrote, 'If every tool, when ordered, or even of its own accord, could do the work that befits it ... then there would be no need either of apprentices for the master workers or of slaves for the lords.'[22] However, the research to develop technologies that replicated human powers was often intertwined with war. Scientists finally managed to create machines that could be controlled from afar and move about on their own, and in the twentieth century (the start of the robotic age) the link between robots and war became even closer. Thus, robotics altered the lethality of war, but also 'the very identity of who fights it'.[23]

The Pioneer, an unmanned plane that was bought from Israel, was one of the first drones to be used by the US Navy in wars from 1986. Pioneers took off battleships and flew over targets to see where shells fired from the ships were landing. First used in the first Gulf War against the Iraqis, they were known in Iraq as 'vultures'. 'When they heard the buzz of a Pioneer overhead, all heck would break loose shortly thereafter because these 16-inch rounds would start landing all around them', said Steve Reid, an executive at the Pioneer's maker, A AI.[24] In one case, a group of Iraqi soldiers saw a Pioneer flying overhead and, rather than wait to be blown up by a 2,000-pound cannon shell, waved white bedsheets and undershirts at the drone.

A crucial year in the history of drones and war was 1995, when unmanned systems were integrated with the GPS – the Global Positioning System – a satellite-based radionavigation system owned by the US government and operated by the US Space Force. The GPS is a constellation of military satellites that provide location and direction of a receiver, anywhere in the world. It allowed unmanned systems and their operators to automatically know where they were at any time. With GPS and the advance of the video-game industry, the interfaces became accessible to a wider set of users. It was drones like Predator and Global Hawk that made their debut in the Balkan wars, specifically in the war against Serbia.

The more the military used unmanned systems, the more people came to believe that machines brought many advantages to the battlefield, beyond their killing capacity. 'They don't get hungry', says Gordon Johnson of the Pentagon's Joint Forces Command.[25] 'They're not afraid. They don't forget their orders. They don't care if the guy next to them has just been shot. Will they do a better job than humans? Yes.' Robots also proved attractive for roles that fill what people in the field call the 'Three Ds': dull, dirty, dangerous. Military missions can be dull as well as exhausting for humans, but drones are able to stay in flight for as long as twenty hours, whereas humans lose effectiveness after ten to twelve hours. The impact on humans is both physical and psychological when they have to carry out the same task over a long period. Unmanned systems, in contrast, do not need to sleep, or eat; they do not get bored or lose focus; they do not need breaks to rest, or to regain concentration. Robots can do the same tasks, in less time and with greater accuracy than human beings. That is partly because drones do not carry 'human baggage'. Unmanned systems can also operate in dirty environments, such as a battle zone filled with biological or chemical weapons, where a human would have to wear protective gear. The drone does not have the same limitations as a human body:

> It used to be that when planes made high-speed turns or accelerations, the same gravitational pressures (g-forces) that knocked the human pilot out would also tear the plane apart. But now, as one study described of the F-16, the machines are pushing far ahead. 'The airplane was too good. In fact, it was better than its pilots in one crucial way: It would manoeuvre so fast and hard that its pilots blacked out.'[26]

How does one intervene safely in places as inhospitable as irradiated zones, in the depths of the sea, or on distant planets? In 1964, the engineer John W. Clark produced a study on remote control in hostile environments:

> When plans are being made for operations in these environments it is usual to consider only two possibilities: either placing a machine in the environment or placing a protected man there. A third possibility, however, would in many cases give more satisfactory results than either of the others. This possibility employs a vehicle operating in the hostile environment under remote control by a man in a safe environment.[27]

As drone technology advances, the human is becoming the weakest link in war. Unmanned systems are providing a path around human limitations, by flying faster and longer, manoeuvring more sharply and generally outdoing the pilot physically and psychologically. In war, working at digital speed is a great advantage over the enemy. Recognising danger and reacting to it before you become its target may be what decides the victor in the end:

Automobile crash avoidance technologies illustrate that a digital system can recognise a danger and react in about the same time that the human driver can only get to mid-curse word. Military analysts see the same thing happening in war, where bullets or even computer-guided missiles come in at Mach speed and defenses must be able to react against them even quicker.[28]

Mach Speed is when an object moves faster than the speed of sound. Mach 1 is the point at which an aircraft exceeds the speed of sound, creating a sonic boom; at Mach 2, the aircraft is travelling twice the speed of sound; and Mach 3 means an aircraft is travelling three times the speed of sound, and so on.[29] Its name is in honour of Ernst Mach, the first physicist to use this unit.

A great feature of unmanned vehicles is their relative independence or autonomy from humans and all their weaknesses/vulnerabilities. There are, however, variations in the autonomy of drones, from direct human operation to the machine becoming 'adaptive'. Singer provides this list:

- 'Direct human operation' is having someone behind a computer control all the plane's operations from the ground.
- 'Human-assisted' is when the pilot on the ground takes off and lands the plane but can let the plane fly itself.
- In 'human delegation', the pilot instructs the plane to take off and land and gives it waypoints to fly to.
- In 'human-supervised', the operator is no longer a pilot, but simply monitors the information that the drone sends back.
- In 'mixed-initiative', the human gives the drone a mission to accomplish, but does not need to oversee it.
- In 'fully autonomous' mode, the machine decides on its own what to report and where to go.
- In its most advanced state, the machine is 'adaptive' when it can learn; it can update or change what it should search out. This last state is the ultimate goal of drone warfare, as it would remove all human agency and render the drone war perfect in terms of efficiency and ethics. The machine will have learnt how to make the perfect – most accurate and fairest – strike.

Autonomy is not only about whether the human is in control or not, or how much in control, but also about how the machine relates to the world. In the context of war, it is easy to see the attraction of increasing the levels of autonomy of military robots. The more autonomy they have, the less support they need. The less support they need, the less likely they are to err, since error is associated with humans and their shortcomings. With the rise of more sophisticated sensors, faster computers and GPS that can give location and destination simultaneously, higher levels of autonomy are becoming more attainable and cheaper.

Warfare without risk is what drone warfare promises, first and foremost, with aircrew who never leave the ground, who are unseen, and who operate at the cutting edge of technology. The 'man in the safe environment' is a telechirist; the drone, a telechiric machine. The machine may be thought of as an alter ego for the man who operates it, argues Chamayou: 'His consciousness is transferred to an invulnerable mechanical body with which he is able to manipulate tools or equipment almost as though he were holding them in his own hands.'[30] This device implies a specific topography, way of thinking and organising space: space is divided into a hostile area and a safe one. One intervenes in the hostile zone and patrols it but does not go there, 'except to carve out new secured zones, bases, or platforms in accordance with a general topographical schema and for reasons of security'.[31]

The strategy of militarised manhunting is essentially preventive; what Chamayou terms 'prophylactic elimination'. When it works best, it is not just about responding to actual attacks, but also about preventing the development of emerging threats by the early elimination of their potential agents. The political rationale that underlies it is that of social defence; its classic instrument is the security measure that is not designed to punish, but only to preserve society from the danger that dangerous individuals within it present. As such, it can even be termed and celebrated as a humanitarian weapon.

Drones and Ethics

The drone has been called a humanitarian weapon, as well as a moral one. The hunter-killer drone, writes Anderson, is 'a major step forward in humanitarian technology'.[32] The drone, it is argued, is humanitarian as a weapon, as a means of killing, for several reasons. First and foremost, nobody dies, except the enemy: the bad, dangerous people, who are causing harm and suffering. Taking lives without even endangering the lives of one's own people (civilians and 'combatants') is good. In effect, the drone saves our lives. According to Strawser, the drone is not just morally acceptable, but morally obligatory. If you want to kill morally, you must use a drone. He argues based on the principle of unnecessary risk, according to which it is wrong to command someone to take on unnecessary potentially lethal risk. With drones, the old military dream of surgical strikes becomes a reality, and necroethics takes the form of a doctrine of killing well.

It appears that technical progress and new weapons have made it possible to achieve risk-free warfare: for our soldiers and for their civilians. The new technology has eliminated moral concerns and dilemmas. Drones 'have the potential for tremendous moral improvement over the aerial bombardments of earlier eras'.[33] A CIA agent expands on Strawser: 'Look at the firebombing of Dresden and compare what we're doing today.'[34] Certainly, compared to a Second World War bomber, 'the drone undeniably gains in precision',[35] but does it enable 'good' wars?

Remoteness and Easy Killing

Compared to the bomber, the drone indeed spares (not saves!) lives. Yet while lives may be spared, unmanned systems change the way people perceive war. This affects a war's processes and outcomes. Unmanned systems lower the threshold for going to war and make it more acceptable; wars with fewer casualties (or no casualties at all for 'our' side) seduce us into more wars. As for the public, it also becomes distant from and disinvested in its foreign policy, as unmanned technologies cut off the last remaining threads of connection between military and society. War is experienced differently both by citizens and by cubicle warriors. Singer calls it 'war via chat rooms' – less like band of brothers and more like Facebook friendships.

> Participation via the virtual world also seems to affect not merely how people look at the target, but also how the person looks at themselves (why people in online communities, for example, take on identities and try out behavior they would never do in real life, be it wearing tattoos

or sharing intimate personal information with strangers). Research shows that this sort of 'externalization' allows something called 'doubling'. Otherwise nice and normal people create psychic doubles that carry out sometimes terrible acts that their normal identity would never do.[36]

As technology moves soldiers further and further away from their enemies, battle becomes more and more impersonalised. Killing from a distance means clicking the apparatus, and when you click, you kill. The act of killing is reduced to positioning the pointer or arrow on little 'actionable images', tiny, blurred figures that have replaced the old flesh-and-blood body of the enemy. What the telechirist sees are little figures blurred into facelessness. A former CIA officer declares, 'You could see these little figures scurrying and the explosion going off and when the smoke cleared, there was just rubble and charred stuff.'[37] Reducing flesh-and-blood humans to tiny, blurry images on a screen helps to make the homicide easier. In such killings, there is a) no physical soiling; b) psychological ease; and c) a (false) sense of the absence of moral soiling. Psychologist Stanley Milgram suggested that 'possibly it is easier to harm a person when he is unable to observe our actions than when he can see what we are doing'.[38] The agent is not only spared the physical and psychological soiling that comes with remote killing, but also spared the shame that may be prompted by seeing one's actions with the eyes of one's victims. He or she will never see the victim looking back at them, as they prepare to kill him. The psychic discomfort is further mitigated by what Milgram called the break in the 'phenomenological unity' of the act. By pressing a button here, some silhouettes representing the bad people disappear in an explosion over there. 'There is a physical and spatial separation of the act and its consequences', Milgram writes. 'The two events are in correlation, yet they lack a compelling phenomenological unity. The structure of a meaningful act – hurting a man – breaks down because of the spatial arrangements.'[39] 'The filtered nature of perception, the figurative reduction of the enemy, the nonreciprocity of the fields of perception, and the dislocation of the phenomenological unity of the action are all factors that ... produce a strong buffering effect', argues Chamayou.[40] The drone presents its operators with powerful means of distancing: the violence of warfare is being exercised from a peaceful zone without risk. The war becomes one-sided and more like a militarised manhunt; global, lethal policing. The armed Predator's main strategic role, for example, is to conduct strikes on high-value targets. These strikes are sometimes called targeted killings and often take place in Pakistan, Afghanistan, Yemen and Iraq. Ultimately, warfare takes the form of vast and long (perhaps endless) campaigns of extrajudicial executions.

The alienating effect of distance has led to concerns that these are 'video-game wars', where figures on a screen are killed almost as recreation, or with the ease of recreational pretend-killing, as drone operators describe their living, breathing human victims as 'bugsplat'.[41] In the twenty-first century, the drone wars have led to the emergence of a new vocabulary of war, a vocabulary that includes the term 'PlayStation killers' to refer to the crew who operate them. With the advances in drone technology war and warfare have become distant, less problematic and easy, and anything that makes it ethically easier to wage war and easier to take life cannot be a good thing.

Intelligence and Emotions

Computers outperform us in tasks that involve numbers, calculations and searching for stored information. They can remember 'trillions of points of data, whereas most of us have a hard time remembering even the PIN number to out ATM at the bank'.[42] In every way, they are more intelligent than us humans, more logical, have better-functioning and more reliable 'brains'. Their intelligence, however, is limited to the world of numbers, and sometimes 'this logical, mechanical manner of intelligence is hard to translate to activities in the real world'. Moreover, intelligence is also the ability to act appropriately in an uncertain environment. How does a machine judge an appropriate act or response, especially in the chaos of war? While it may be morally right to remove at least some humans from situations of risk and danger, their mechanical intelligence coupled with their (varying degrees of) autonomy increase the risks surrounding their moral choices. Technology has often moved faster than the laws of war, and we are now faced with the possibility of robots making life-or-death decisions.

War is strategy and policy, calculating, planning and executing well-considered political aims, but it is also instinctive, emotional and personal. It inevitably entails victories and defeats, joyful triumphs and crushing losses, pride in accomplishments and despair in failures. For those who fight, war means the risk of death or injury, of oneself as well as of comrades. As a result, soldiers and their families, the nation itself, all care about a war that is being planned or fought. War is fuelled and sustained by human emotions: fear, anger, honour, pride, the drive to survive, but also the desire to sacrifice oneself. On the negative side, such emotions can lead to war crimes, as soldiers are often overcome by 'the heat of battle': horrific conditions, human losses, witnessing terror and grisly killings. Such experiences can lead to acts of vengeance and brutality. Technology is one way to reduce war's costs and passions, and therefore its crimes. Robots have no emotions. Unmanned systems remove the anger and other negative emotions that can be generated within a bloody battlefield. A remote drone operator is not a combatant and is not experiencing emotions associated with the prospect of being killed or the experience of watching others being killed around him. This lessens the likelihood of civilians being killed, making the war more ethical. However, it also makes 'soldiers' too cool, or not much affected by a kind of killing that consists of eliminating icons on a computer screen – yet it remains the killing of human beings. With emotions removed, the resistance to killing may, in fact, be much lower, because it does not seem or feel like killing, even though it is.

As robots gain more and more autonomy, emotions will be completely removed from the equation. A computer has no anger, but it also has no pity, no conscience, no mercy and no sense of guilt. In addition, it has no pride, honour or courage:

> The attempt to eradicate all direct reciprocity in any exposure to hostile violence transforms not only the material conduct of armed violence technically, tactically, and psychically, but also the traditional principles of a military ethos officially based on bravery and a sense of sacrifice. A drone looks like the weapon of cowards.[43]

Its supporters, however, declare it the most ethical weapon ever known to humankind, the humanitarian weapon *par excellence*.

It could be argued that one ethic is being substituted for another, as technology has been replacing the military ethic of self-sacrifice and courage with one of selfishness, self-preservation and what amounts to cowardice. It is said that killer-drone operators are rather detached, disengaged, remote and emotionally disconnected from their targets/victims, and that it is a good thing that they are thus protected. It may indeed be good for them. Increasing drone autonomy would make the killing by machines completely emotionless. However, that would not make those cold-blooded killings moral, noble or humane.

Immorality

Drone wars are wars without human costs, but remote and riskless warfare can undermine the morality of even 'good' military campaigns. When a nation decides to go to war, the very decision itself is a reflection of the moral character of its people. This technology disengages the public from foreign and defence policy and turns the real-life killing of people into something that is watched, even as entertainment. Moreover, the usual checks and balances that provide the basis for democracy are suspended, which jeopardises the idea of a democratic peace. Without public debate and support, and without risking troops, the decision of war may reflect a nation that just does not care. Even if the nation acts on a just cause, war can be viewed as merely an act of selfish charity:

> One side has the wealth to afford high technologies and the other does not. The only message of 'moral character' a nation sends out is that it alone gets the right to stop bad things, but only at the time and place of its choosing, and most important, only if the costs are low enough. While the people on the ground being saved may well be grateful, even they will see a crude calculation taking place that cheapens their lives.[44]

With unmanned systems even the best wars lose their virtue and become like playing God from a faraway, high place, using violence and taking lives with impunity. Attention is no longer paid to the numerous ethical questions that all wars raise, which inevitably leads to breaches of human rights and to acts of immorality. Since in war killing is not only allowed, but is actively encouraged, even considered one's duty, war can easily become a series of killings, destruction and devastation, not the rule-governed activity of equals who have the same human standing that international law commits to states, nations and individuals. New war technologies create new challenges for international law and for traditional war ethics.

As a powerful nation projects power through the use of its technology, what is obscured is the wounding and killing, the destruction of homes, schools and villages, the displacement of frightened families, terrorised by the presence of lethal mechanical assassins. Projecting power without projecting vulnerability implies that the only vulnerability will be that of the 'other', who becomes a mere target. Killing and injuring occur only in one direction; people still die, in some cases in their thousands, but only on one side. For those who use such weapons, it becomes *a priori* impossible to die as they kill. From being symmetrical or asymmetrical, conventional or unconventional, regular or irregular, warfare quite simply and absolutely becomes unilateral. As for combat, it becomes a campaign of slaughter. The aim is to kill, rather than capture; to take lives, rather than spare, protect, respect another person's right to life – the most basic human right.

What about the drone's capacity to discriminate in the choice of targets? 'The fact that your weapon enables you to destroy precisely whomever you wish does not mean that you are more capable of making out who is and who is not a legitimate target', writes Chamayou.[45] In practice, their persistent surveillance is not particularly good when it comes to drawing distinctions between legitimate targets and protected persons. But does this matter? If the drone is represented as being virtuous, that is because it makes it possible to rule out any possibility of casualties in one's own camp. The argument was summed up in a 2011 British report: 'the use of unmanned aircraft prevents the potential loss of aircrew lives and is thus in itself morally justified'.[46] However, the law on armed conflict limits the exercise of armed violence, because it is based on the universal rights of human beings, whoever they are: one's co-nationals or foreigners. When there is such clear and self-admitted separation between humans that are 'ours' and those that are not (and, as such, are less deserving of the right to live), we see nationalism at its extreme and at its most immoral. 'According to our norm of priorities on grounds of duties, the state should give priority to saving the life of a single citizen, even if the collateral damage caused in the course of protecting that citizen is much higher', write Kasher and Yadlin.[47] Just war's *jus in bello* distinction and proportionality are sacrificed in the name of saving national lives. Just like in the Second World War bomber wars, the safety of our soldiers takes precedence over the safety of their civilians.

Regarding the alleged precision of killer drones (which will be further explored in Chapter 4), there is a crucial difference between hitting the target and hitting only the target. The AGM-114 Hellfire fired by the Predator drone has a kill zone of 15 metres. What this means is that anyone who happens to be within a 15-metre radius around the point of impact, even if they are not the designated target, will die together with the target. As for the wound radius, it is estimated to be 20 metres.

The number of civilians killed by drone strikes has been a matter of debate, especially in the past ten years. An investigation by the *New York Times* found that in official reports, all military-age males killed in strikes were considered to be combatants 'unless there is explicit intelligence posthumously proving them innocent'. Estimates from 2014 of the number of civilians killed in Pakistan range from fewer than 150 to almost 1,000, out of approximately 3,000 killed.[48]

In effect, drone warfare does away with combat and, in so doing, calls into question the very notion of war. This gives rise to an important question: if drone wars are not, strictly speaking, warfare, then what kind of 'state of violence' do they amount to? A number of possibilities come to mind, all lacking in morality:

- far-flung imperial conquests;
- the remote-controlled hunting of human beings; and
- permanent lethal surveillance and annihilation.

The worst scenario may be that drone wars are wars without victory and without end: infinite state violence, with no limits or possible exit.

Drone warfare leads to a crisis in military ethos. The cardinal virtues, the four virtues of mind and character in classical philosophy and Christian theology – prudence, justice, fortitude and temperance – form a virtue theory of ethics. Military morality has its own, very similar, cardinal virtues: courage, justice, sacrifice, heroism, wisdom, integrity, temperance, honour, perseverance, and so on. Those values have always had a clear ideological function: to make the butchery in war acceptable – even glorious. Being 'ready to die' seemed to be one of the main factors in victory, the very heart of what Clausewitz called 'moral strength'. War was the moral experience *par excellence*. However, warfare dictated by the imperative of self-preservation makes heroism, courage and other traditional military virtues impossible. Killing by drones emerges as a virtueless war of cowardice and dishonour in a post-heroic age. When there are humans behind the killer drones, those humans are simply assassins. Even when fighting a (just) war of defence, the attacked state can use any means of defence, wrote Kant in the *Doctrine of Right*, 'except those whose use would render its subjects unfit to be citizens ... It must accordingly be prohibited for a state to use its own subjects ... as poisoners or assassins.'[49] The principle that Kant formulates concerns what a state may not make its citizens do. What a state can make its citizens do is limited by what that would make them become. The principle of citizenship forbids the state from ordering its soldiers to assassinate an enemy, employing weapons that deprive the enemy of any chance of fighting back. Kant declares that a state does not have the right to turn its own citizens into assassins.

Drone strikes have other effects too. The 'Living under drones' report by Stanford International Human Rights and Conflict Resolution Clinic suggests that the drone war in Pakistan is having long-term impact on the mental health of those who live with it. According to the report, drones cause 'constant and severe fear, anxiety and stress' to people who cannot protect themselves, and who may develop phobias about social gatherings, attending school, or even going to the market, as a result of witnessing missile strikes or their aftermath.[50]

All that terror and destruction was justified by the alleged effectiveness of drones. The strikes have been highly effective at taking out specific named targets, and there is no doubt that the leadership of Al-Qaeda and other groups has been decimated by them, but destroying the leadership does not bring victory; a leaked 2009 CIA report on targeted killings (released by Wikileaks in 2014) noted that they had limited effect against the Taliban in Afghanistan because leaders were replaced rapidly. The 2009 report analysed 'high-value targeting' (HVT) – the assassination of senior insurgents – in a number of conflicts. The report, which Wikileaks described as 'pro-assassination', looked at the pros and cons of HVT programmes.[51] As well as examining recent actions in

Iraq and Afghanistan, the report assessed British action in Northern Ireland, Sri Lankan operations against Tamil Tigers and French efforts during the Algerian civil war, among others. Benefits of HVT operations, according to the report, include eroding insurgent effectiveness, weakening insurgent will and reducing the level of insurgent support, while potential negatives include strengthening an armed group's bond with the population and radicalising an insurgent group's remaining leaders. Following the publication of the CIA report in July 2009, President Obama increased drone strikes in north-west Pakistan, in Yemen and, five years later, in Iraq. The use of drones against high-ranking individuals has also been criticised for reducing the chances of any negotiations with the groups that are targeted.

Drone Wars from Three Perspectives

Militarism

Militarists would be critical of drone warfare for precisely the reasons why such warfare is considered moral. To begin, for militarists total war – what drones are there to do away with – is not a problem. War is a permanent human obligation and total war may be necessary for the moral army to defeat existential threats to the nation, to humanity, to some great and valuable ideal. In total war the moral army's strengths and virtues will be revealed and honoured. War may even bring great social change, which will be the aim or the good cause behind it. The evil enemy will be defeated best and completely through total war, as moral rewards and punishments will be meted out during its course.

For militarists, drone war is a war of cowards. 'War is the romance of history. Inordinate ambitions are the soul of every patriotism, and the possibility of violent death the soul of all romance', writes William James.[52] It is a sort of sacrament. Heroism, self-sacrifice, courage, all that military ethos is removed from war and what remains is war without victory, without glory. The moral strength of soldiers being 'ready to die' was what made war the moral experience *par excellence*. In drone wars, the only ones who are 'ready to die' are the enemies – who emerge as the only heroes.

Militarism is the great preserver of our ideals of hardihood, and human life with no use for hardihood would be contemptible. 'All the qualities of a man acquire dignity when he knows that the service of the collectivity that owns him needs them', writes James, and his own pride rises. 'No collectivity is like an army for nourishing such pride.'[53] Martial virtues are the enduring cement of nations and permanent human goods; they are patriotic pride and ambition in their military form, at the heart of all nationalisms. Modern man inherits all the innate pugnacity and all the love of glory of his ancestors. War is the strong life; it is life *in extremis*. James writes that martial virtues like 'intrepidity, contempt of softness, surrender of private interest, obedience to command must still remain the rock upon which states are built'.[54]

In remote warfare the military – the nation's greatest pride – is disconnected not only from the war, but also from society and so from the nation itself and from all the nation's myths: the myth of the homeland (the land that 'belong to us'), the myth of ethnogenesis (stories about 'our glorious ancestors'), the myth of election (the values that make 'our nation' special) and the myth of suffering (denoting who 'our enemies' are). All those myths construct and sustain the nation's identity and mission through the ages. They bind communities, create emotions and can move and mobilise the entire nation in a time of crisis. Nationalism – a celebration of the nation – is both a 'collective sentiment or identity bounding and binding together those individuals who share a sense of large-scale political solidarity'.[55] National myths of territory, suffering and redemption, unjust treatment, military valour, kinship and shared cultural values, address deep psychological needs that must be satisfied for society to hold together and for the nation to survive, in the context of a specific accepted past and a future destiny. In this way, ethics, nationalism and militarism are necessarily interlinked, through the moral chain of defence of a glorious nation and its sacrifices. The use of unmanned systems undermines all elements of nationalism – the stories of past glory, the special moral character of civilians/nationals, the noble mission of defending the patria and the perceptions of its warriors – as it is from suffering and fortitude that nations are able to draw strength and pride and survive.

The connections between nationalism and military valour, being 'ready to die' for the homeland and honouring one's heroic ancestors, can be seen in the national anthems that tell of a glorious past and a glorious present, filled with bravery, virtue and self-sacrifice, usually emerging through wars. What follows are some examples.

Russia

> Russia is our sacred state,
> Russia is our beloved land.
> The powerful strength and the great glory
> Are your properties for all the time.
> Long live our Fatherland, land of the free,
> The eternal union of brothers nations,
> Given by the wisdom of ancestors
> Long live our land! We are proud of you!

Italy

> Italy has awoken with Scipio's helmet
> ...
> We are ready to die.

Scipio Africanus was a Roman general and statesman, regarded as one of the greatest military commanders and strategists of all time.

Germany

> German fatherland
> Inspires to noble deeds.

Turkey

> I am ready to sacrifice myself for you
> ...
> Smile upon my heroic nation
> If you frown our bloodshed for you will not be worthy.

Iraq (since 2003)

> Oh my homeland, the youth will not tire
> Their goal is your independence
> Or they perish

> We will drink from the cup of death
> And will not be slaves to our enemies.

Iraq (1981–2003)

> We pledge by sword and the speech of pride
> And the neighing of horses when in duty
> We are the wall of its limitless range
> And the roarings of people in days of war
> We inherited the flags of the Prophet from desert
> ...
> Oh expanse of glory, we have returned anew
> To a nation what we build with unyielding determination
> And each martyr follows the footsteps of a former martyr
> Our mighty nation is filled with pride and vigour.

The American national anthem contains the words '*gallantry*' and '*bravery*', and tells of a '*perilous fight*', the '*havoc of war and the battle's confusion*'. The Greek anthem is to liberty, which is seen '*in the cut of the sword*' and emerges '*from the sacred bones of the Greeks*' with '*ancient valour rising*'. The national anthem of the United Kingdom asks that the King be '*saved*' (originating from back when Kings fought in battles), and declares the King '*gracious*', '*noble*' and '*glorious*'. The Irish national anthem is '*A Soldier's Song*'.

Drone systems objectify and dehumanise both the people targeted by them and those who operate them. What passes as virtuous and noble, the computerised postmodern warfare, where drone operators and the public view war from a distance, through the mediation of weapons systems, computer screens and television screens, is nothing of the sort. If anything, for militarists, it takes away all moral value that is necessarily part of a good war, both for the good soldier and for the evil enemy. Instead, what happens is that the 'soldier' appears to achieve a moral dissociation from the targeted 'things' on the screen; this is not a moral relationship. The technological mediation vital to 'the drone stare' is framed by advocates of unmanned systems as an unproblematic ability 'to see the truth of a particular situation or to achieve a totalizing view of the "object" under cosmic control'.[56] In the words of Robins and Levidow, 'Enemy threats – real or imaginary, human or machine

– became precise grid locations, abstracted from their human context.'[57] This is all unacceptable to the ethics of militarism.

Realism

Realists would be in support of unmanned system, mainly because they preserve military lives, as well as the army's reputation. Unlike bomber strikes, drone strikes keep the overall casualty count low and so spare lives on both sides, thus mitigating actual physical harm and protecting our armies and nations from accusations of war crimes. This theory sees war as inevitable, because fighting is a fundamental tendency in human beings and also because it is through war that states gain power and ensure their security.

> War – or organised fighting between large groups of adult human beings – must be regarded as one species of a larger genus, the genus of *fighting*. Fighting is plainly a common, indeed a universal, form of human behavior ... Wars between groups within the nation and between nations are obvious and important examples of this type of behavior ... The simplest and most general causes of war are only to be found in the causes of fighting.[58]

As fighting, selfishness and greed are part of human nature, aggression at all levels – individual, collective, internal and external – is inevitable:

> Whether one looks back through time or downward to simpler forms of social organization, it is a common practice for individuals or groups to seek to change their environment by force and for other individuals and groups to meet force with force.[59]

Possessiveness is a common cause of fighting: to establish and defend exclusive rights to possessions. Those necessary and inevitable drone wars can provide physical, psychological and moral buffering. They are powerful weapons at the command of powerful states (which realism favours) that help ensure control, survival and security. The more effective they are in their targeted assassinations, the better. Not that realism is opposed to total war resulting in thousands of casualties; after all, any casualties on the enemy side would not be regarded as moral crimes by realists, because war is amoral. However, if casualties can be kept low, if all war costs can be kept to a minimum, the state gains in power and in security. Moreover, the state becomes feared, rising in status and prestige, because it can easily start or resume those remote wars anytime and anywhere it detects threats to its interests.

The surgical strike from a space of safety, clean and clinical, devoid of all emotion and moral value, is a realist's dream. Precision killing, the practical action of firing a Hellfire missile on human beings, aimed at taking their lives, is translated and transformed by the informational system into a computerised checklist of 'things to do'.

Just War

It is the just war perspective that presents us with a more complex picture, which ranges from condemnation to acceptance.

The question of peace is an important one. Grey zone conflict means potentially endless war; as much as just war accepts that war is sometimes necessary, it is in support of the humanity and the morality of peace, where those who violate human rights and commit crimes such as murder and the destruction of property are held accountable. The lack of justice and accountability is tightly linked to the terrifying prospect of endless war, because drone warfare involves violence used with impunity, which is unjust and immoral; and the likelihood of the permanence of this state of unpunished cruelty enforced on human beings. Can there be a just and peaceful world under the threat of drone attacks? What drone wars – as all and any wars – still cause is human suffering and the prospect of autonomous machines making (what should be) ethical decisions.

Just war would also be critical of remote killing for being easy, video-game killing, where a person is depersonalised, dehumanised and simply a blurry target on someone's screen. The drone cannot know who that person is, not with any accuracy, before the strike, and sometimes even after. It cannot even know for certain if the human is a civilian, or a combatant. The policy of kill, rather than capture, is what adds to this violation of human rights, assaulting life and dignity.

A one-sided combat is not combat, it is assassination. Just war would consider unethical the assassination of individuals that have no chance of protection, no possibility of fighting back or of surrender. Such endless lethal aggression does not fit well with the morality of just war. But does it matter who dies? Let us examine the so-called 'precision strikes' by looking at two examples: one where the strike was not precise and one where it hit the intended target.

Case 1 – Kabul, 29 August 2021: The Killing of Ten Civilians

The day of the drone strike, Afghan aid worker Zemari Ahmadi ran errands for his employer, which included picking up a laptop and delivering water. Mr Zemari, employed as an engineer at the US-based non-profit NEI foundation, dropped off some of his colleagues and headed home to an evening meal with his family – a drive he did routinely. For the Ahmadis, that Sunday afternoon was like any other. Cousins, friends and families of different age groups were mingling in the family courtyard. Mistaking the compound for an ISIS safe house, six Reaper drones surrounded it and, despite the clear presence of civilians, at 4:53 pm, a single Hellfire missile was launched, killing seven children and three adults.[60] The attack was conducted by the Over-the-Horizon (OTH) Strike Cell group of the US Central Command.

Those killed were Zemari Ahmadi (aged forty), his son Zamir (aged twenty), Mr Ahmadi's cousin Naser (aged thirty) and the following children: Faisal (aged sixteen), Farzad (aged ten), Arwin (aged seven), Benyamin (aged six), Malika (aged three), Somaya (aged three) and Hayat (aged two).

This was a deliberate attack on civilians, who had been visibly moving around the car that was struck, as the release of the video under the Freedom of Information Act made clear. Even if the man inside the car had been the correct target – that is, he was ISIS affiliated – the missile was still launched in the knowledge that several civilians in the vicinity of the target would be hit and killed as well. This incident casts doubt on the alleged surgical precision of the drones, but more importantly stresses the ethical implications of the risky environment created by drone warfare for abuse and systematic inhumaneness. There have been no arrests or any accountability, following the mass killing of ten innocent people who, in theory, are morally and legally 'protected'.

In the best possible scenario, this was gross negligence.
Gross or culpable negligence arises when the person:

i. is expected or required to abide by certain standards of conduct or take certain specific precautions; and
ii. is aware of the risk of harm and nevertheless takes it, believing that the risk will not materialise owing to the steps that he or she has taken or will take.[61]

Even as such, it was wrongdoing and culpable.

In the debates about the principle of necessity, there are questions that must be answered, in this order:

1. Is there a serious harm that should be prevented?
2. Is the harm imminent or extremely likely to occur?
3. Is there no other, less problematical way to prevent the harm?
4. Does the proposed means cause less harm than it prevented?

Positive answers should be given to all of the above.

Regarding question 1, in weighing harms, it is not enough to justify the use of a problematic tactic that there is some minor advantage and no overriding disadvantage in using the tactic:

> Military necessity concerns the justifications of tactics that would normally not be considered justified. And because the normative considerations that are usually cited for initially ruling out that tactic are serious, a similarly serious consideration is needed now to rule them out.[62]

Regarding question 2, the requirement of imminence, there must be a preponderance of evidence that serious harm will occur – and soon – unless the tactic is employed. For example, instant and overwhelming necessity for self-defence. Imminence is supposed to signal an emergency. Regarding question 3, are there any other avenues through which the harm can be prevented? That is, avenues that are legal or at least less problematic? The invocation of military necessity must not become a way of violating basic human rights. It is 'not a blanket justification that can transform any rule into one that admits of an endless class of exception'.[63]

In question 4, proportionality enters the picture to urge that only the strategy that causes the least suffering is chosen. The tactics employed must not cause more suffering than they prevent. If there are other means available, the principle of proportionality forbids the use of the more costly tactic. When necessity is combined with proportionality, even fewer cases will be justified. For even when attacking a legitimate military target, disproportionate use of force is morally unacceptable. Humane treatment (a moral notion) is at the core of the principle of proportionality.

What this case shows is not only lack of surgical precision on the part of drones, but also indiscriminate treatment, superfluous suffering, lack of compassion and mercy, and subsequent lack of accountability – all of which go against the principles of just war.

Case 2 – Baghdad, 3 January 2020: The Killing of Iranian General Qasem Soleimani and Nine Paramilitaries

Qasem Soleimani was one of the most powerful figures in Iran, after its Supreme Leader. As commander of the Revolutionary Guards' overseas operations arm, the Quds Force, he was 'an architect of Iranian policy across the Middle East'.[64] Being in charge of Quds meant that he secured funding and weapons, and provided logistical support to allied governments and armed groups, including Hezbollah, Hamas and Islamic Jihad. US President Trump claimed that under Soleimani's leadership the Quds Force had 'targeted, injured, and murdered hundreds of American civilians and servicemen' over the past twenty years, and that he had orchestrated a rocket attack in Iraq in December 2019 that had killed an American contractor. The general arrived at Baghdad International Airport on a flight from Syria early on 3 January. He was leaving the airport with senior officials from Iraqi Shia militias when their convoy was hit by missiles fired by a US drone. Among the people killed alongside Soleimani were Abu Mahdi al-Muhandis, deputy commander of Iraq's paramilitary Popular Mobilisation Forces (an umbrella grouping of Shia militias) and eight other members of Popular Mobilisation.

In this case, the remote strike was indeed precise and hit only the intended target/s. No civilians were hurt. Yet a report by the UN's special rapporteur on extrajudicial killings, Agnes Callamard, declared it an 'arbitrary killing', on the grounds that the United States had not provided sufficient evidence of an imminent threat to life to justify the attack. Consequently, the United States was responsible under international human rights law, according to the report. Callamard presented her report at the Human Rights Council in Geneva on 9 July 2020. In her view, the consequences of targeted killings by armed drones have been neglected by nation states. Her report said that the world is at a 'critical time and a possible tipping point' when it comes the use of UAVs:

> States who used them on the grounds of self-defence, defined in a very elastic fashion against purported terrorists, risked creating a situation where there will be no red lines really, she told journalists ... As more Government and non-State actors acquire armed drones and use them for targeted killing, there is a clear danger that war will come to be seen as normal rather than the opposite of peace ... War is at risk of being normalized as a necessary companion to peace, and not its opposite. Appealing for greater regulation of the weapons, and lending her support to calls for a UN-led forum to discuss the deployment of drones specifically, the Special Rapporteur insisted that their growing use increased the danger of a 'global conflagration'.[65]

Using drones, states may opt to strategically eliminate high-ranking military officials outside the context of a known war and try to justify the killing on the grounds of necessity, not imminence, classifying the target a potential, future threat. Callamard stressed that the drone strike in Iraq that killed Soleimani was a violation of the UN Charter. Targeted killings until January 2020 had been limited to non-state actors, but in this case a state armed drone targeted a high-level official of a foreign state, on the territory of a third state. Previous UN special rapporteurs on extrajudicial killings have also lamented the lack of clarity among states about their obligations regarding drone warfare and the absence of accountability. The test for anticipatory self-defence is very narrow: it must be a necessity that is instant and overwhelming. This standard, according to Callamard 'is unlikely to be met'.[66]

Back in the 1930s, Weil argued that the most defective method to approach warfare would be 'in terms of the ends pursued and not by the nature of the means employed'.[67] Since then, a number of immoral, cruel, barbaric and inhumane means have been employed in wars around the world. As with other types of warfare, the ethics surrounding precision strikes by drones present us with moral dilemmas and are again incompatible with the principles of just war theory, on which international human rights law is grounded.

Or are they?

Let us look at just war again. First, if the use of drones mitigates harm, then drone warfare is more ethical. Remote warfare using drones – unlike direct combat-type war, or Second World War-type aerial bombardments – does cause fewer casualties. Then comes the principle of discrimination. If artificial intelligence makes it possible to entirely avoid the killing of civilians, then the principle is respected. As a result, by putting our trust in AI, future wars can look a lot less bloody and in agreement with international law, protecting the vulnerable while pursuing the dangerous. Only such force as is morally appropriate may be used, killing only the 'bad' people and not that many of them either – not compared to a state of total war. Just war then comes out in support of drone warfare, at least in principle, that is, if it actually achieves what its proponents claim it is capable of. For that to be possible, two elements are vital: trust and control.

Trust involves an interaction between two or more agents. Trusting requires optimism and reliability, ultimately grounded on the confidence that the trusted party is competent (morally and in terms of ability) to do what they are trusted to do. Traditional definitions of trust assume the existence of a reasonable belief on the part of the trustor in the competency and the goodwill of the trustee. However, trust is also risky and requires that we are vulnerable to others. Balis and O'Neill, from the Royal United Services Institute (RUSI), the United Kingdom's leading defence and security think tank, write:

> What tends to unite most definitions of trust is a sense of vulnerability. Without the possibility of betrayal, without the existence of risk, there can be no trust. It is because of the presumed moral element implied in classical conceptions of trust that some challenge the use of the term to describe the human relationship with artificial agents. At current levels of narrow AI, they argue, we cannot attribute intentionality or moral agency to AI systems, and therefore the use of the term 'trust' is misplaced. Others take a less purist perspective and apply the term in a way that reflects everyday usage implying confidence in the system's reliability.[68]

Still, trust as a term is used widely in computer science and remains a fundamental aspect of public and user acceptance of artificial intelligence. National policies, regulations and expert advice on artificial intelligence today underscore the need for 'trustworthy AI'. For example, the Defense Advanced Research Projects Agency, a research and development agency of the US Department of Defense responsible for the development of emerging technologies for use by the military, has an Air Combat Evolution programme exploring methodologies to model and measure pilot trust in artificial intelligence during air combat manoeuvring.

As technology rapidly advances, positive expectations of the behaviour of an artificial agent may be a sufficient condition for the existence of trust. Trust becomes a facilitator of interactions among the members of a system, whether these be human agents, artificial agents or a combination of both – a hybrid system.[69] At the current level of AI:

> trust appears to be a one-way relationship concerning the degree to which the human 'trusts' the AI, rather than genuinely two-way trust, where the AI takes a view on human performance. Various factors determine (human) trust in technology, including but not limited to the trustor's level of competence and disposition to trust, and the overall environment or context (including broader cultural and institutional dynamics). Beyond such human – and environment specific considerations, what defines the level of trust a person or organisation has in AI are the technology's performance, process (how it generates specific outputs) and, importantly, purpose.[70]

Factors affecting trust in artificial intelligence besides technical ability are fairness, transparency and accountability. This is where the human agency comes in, also described as oversight or meaningful control. Maintaining human oversight of the use of technology may be the only protection against the risk of unintentionally biased, or poorly regulated AI-enabled systems.

Humans will often intervene even in situations where the artificial intelligence is better placed to make decisions and there is no need for control or supervision. Over-trusting is as risky as under-trusting, so a balance between appropriate levels of trust and adequate levels of control in the development and use of artificial intelligence is necessary. This is at the heart of the concepts 'calibrated trust' and 'adaptable/adaptive autonomy': trust is calibrated according to the capabilities of the artificial intelligence, while knowledge and expectations of what artificial intelligence can or cannot do influence the degree of trust. In the case of adaptable autonomy, the user's ability to tailor the level of autonomy can lead to greater levels of trust. This is particularly critical in national security decision-making, especially when it comes to decisions regarding the use of force, where the implications of trusting or not trusting artificial intelligence have the greatest consequences.

The debate over the use of artificial intelligence in supporting military activities is not much different from previous questions regarding the role of technology in human affairs, including its most extreme. It can be argued that this is not a radical departure, but an evolution, requiring familiarisation and increasing confidence over time, as well as continuous efforts to improve efficiency. From a just war perspective, continuous efforts to mitigate civilian harm, with the aims of (a) sparing the vulnerable, and (b) protecting them by targeting the dangerous people (such as terrorists), coupled with increasing trust in AI technology, make remote warfare involving the use of killer drones compatible with its principles.

Conclusion: Liminal Insecurity

Drone warfare is liminal warfare. The term 'liminal' comes from the Latin word for a threshold, and used in the context of war describes the blurred lines or ambiguity that is experienced in grey zone conflict:

> As a form of maneuvre, it is neither fully overt nor truly clandestine; rather, it rides the edge, surfing the threshold of detectability, sometimes subliminal (literally 'below the threshold' of perception), at other times breaking fully into the open to seize and advantage or consolidate gains before adversaries can react. Likewise, the approach exploits undefined or legally ambiguous spaces and categories -using these as cover for action without retaliation.[71]

In the post-Cold War era, liminal warfare took the form of NATO strikes in Yugoslavia 1994–5, by then a failing state, and later in Serbia. It was in the Balkans that Western powers began flexing their military muscle in regions that Russia had considered its turf during the Cold War. To Western publics, the military interventions in Bosnia and Kosovo were presented as necessary, fully justified and moral: humanitarian, altruistic, peacekeeping missions preventing genocide and ethnic cleansing, while simultaneously protecting European security. What was not highlighted was the loss of life directly from those interventions. As the twentieth century was coming to a close, the promises by the UN following the end of the Second World War of a world of peace, prosperity, cooperation, collective security, human rights and the rule of international law, were looking shaky. The development of remote warfare seemed to offer similar promises in areas of liminal wars, ethical interventions and the protection of civilian life. The drone had come a long way since the nineteenth century siege of Venice.

The field of robotics now claims to have delivered on its great promise, most notably through its relationship with the military. When unmanned systems started out, they were seen as strange, they were limited in their use and not quite accepted. As the twenty-first century began, liminal warfare with its liminal ethics was becoming a lot more popular. By 2015, several states had armed drones – the United States, the United Kingdom, Israel, Iran, China, Saudi Arabia, the United Arab Emirates, Pakistan, Iraq, Turkey and Nigeria – establishing an irreversible trend, as a result of their tactical and (alleged) ethical advantages that should satisfy international humanitarian law and, at the same time, eliminate real-world threats. While from a militarist perspective drone warfare ethics (or lack of ethics) are problematic, realism and, ultimately, just war would see a moral and/or legitimate use of killer drones in international conflicts. 'With drones and robots fighting wars instead of citizens, politicians would not need to rally citizens to shed their blood, because they would have a risk-free army to do so',[72] which means that the more autonomous drones become, the more humanitarian they will be. As for civilian casualties, for realism they are simply collateral damage, for militarism, at best, necessary sacrifices to a great moral cause; at worst, members of an evil group/nation/state. Just war proponents would accept some lack of discrimination, as long as the numbers were kept as low as possible and were in proportion. Certainly, just war would favour the further development of killer drones, to make them perfectly precise, so as to remove the possibility of civilian casualties. As for the lack of risk to 'combatants' making war 'cowardly', it can be argued that those who advocate wars of glory, where soldiers gave their lives in bloody battlefields, miss the point that 'we are now in a post-heroic age',[73] where war has become something a lot safer and a lot less romantic (as much as we could argue that the Second World War bombings of entire cities, or the death camps, were 'romantic'). In any case, courage is not the only military virtue. Honour can be shown through respect for *jus in bello* principles, which are not incompatible with the human instinct of self-preservation, and since our species' early development has been our ingenuity to protect ourselves while killing others – even animal much larger and stronger than us.

The following chapter puts those ethics through further tests, as it examines the use of remote warfare in a war of aggression, the war on terror, by the US-UK Coalition, where, it has been argued, state killing enacted by UAV systems exists in a discursive and symbolic context where a belief in or claims to precision technology are used to justify the techno-scientific violence of the West, and countries such as the United States and the United Kingdom embrace this rhetoric to claim that they have been conducting war in more legal and moral ways than less technologically advanced countries.

4 • Remote Killing in the War on Terror

> What resembles 'LITTLE BOY' (one of the atomic bombs dropped on Japan during World War II) and as LITTLE BOY did, represents the dawn of a new era (at least in SIGINT and precision geolocation)?
>
> If you answered a pod mounted on an Unmanned Aerial Vehicle (UAV) that is currently flying in support of the Global War on Terrorism, you would be correct.
> NSA DOCUMENT DATED 3 MARCH 2005,
> RELEASED BY EDWARD SNOWDEN.[1]

While drones were developed and used in the twentieth century, the drone age effectively commenced on 3 November 2002, when the CIA used a Hellfire missile-equipped UAV, the MQ-1 Predator drone, to kill the six occupants of a vehicle in Yemen. Yemeni authorities identified one of the men as Ali Qaed Sunian al-Harithi – 'Abu Ali' – a man believed to be responsible for a terrorist attack on 12 October 2000, when the USS destroyer Cole had a 40-foot-wide hole blown into its side by suicide bombers, an attack that had killed seventeen and injured forty US sailors. When the suicide bombers detonated their explosives, the US destroyer was making a fuel stop at the port of Aden, in Yemen, *en route* to the Persian Gulf. US authorities claimed that all six men in the vehicle were terrorists, members of Al Qaeda, and therefore legitimate targets. The strike was celebrated by the United States and by mainstream media as a triumph of technology, 'the use of a sophisticated machine to terminate with extreme prejudice the enemy while risking no harm to good soldiers fighting a just war',[2] a war that had been termed the 'War on Terror', declared in response to the 11 September 2001 terrorist attacks. From that day, the use of drones to kill suspects/enemies abroad became a frequent, normal, standard operating procedure. Soon, killing remotely stopped being an exceptional measure, but the ethical response to the evil posed by global terrorism. We will never know if those six men in Yemen, or most of those killed by drones, were guilty of terrorist offences. After a drone strike, the vehicles become little more than scrap metal, while the human bodies, after being blown apart, become unidentifiable.

Still, many in the West support the use of drones to 'neutralise' enemies in faraway lands, mostly along utilitarian lines, that is, for 'the greater good'. Sacrificing the few to maximise positive outcomes for the many can be argued to be the right action, the most moral choice. While advocates of the use of armed drones to kill those suspected of terror offences acknowledge that not all suspects – and not all who are killed – are guilty, they still accept, on utilitarian grounds, that it is permissible to kill, injure, or cause suffering to some innocent people, if enough guilty ones are dispatched in the process, making the world a safer place for everyone else.

So, Predator drones can prey on and eliminate those suspected of possibly conspiring to commit terrorist acts sometime in the future. As a method it sounds more than a little suspect itself. 'Remote-control killing' in the context of the War on Terror has been touted as a crucial and moral tool in the necessary response to terrorism, but the cold-blooded, summary execution (without trial) of unarmed people in other states thousands of miles away has allowed for the obliteration of anyone regarded by the killer as worthy of violent, premature death.

The Bush administration notoriously crafted innovative briefs to legalise certain policies:

> Locutions such as 'unlawful combatant' began to pepper official documents, allowing policy advocates to deny the very applicability of the Geneva Conventions (even if valid) to the persons either indefinitely detained or killed by Predator drone. Arguably the most jarring change in policy was the explicit conflation of *offense* and *defense* so as to rationalize proactive or 'preemptive' war.[3]

This new approach was expressed in the *National Security Strategy of the United States of America* issued in September 2002, in which it was stated that the United States considered a good offence the best defence. Targeted killing as a US policy started with George W. Bush, but President Obama, years later, ordered hundreds of drone strikes in at least six countries, surpassing even the self-declared 'War President', famously killing head of Al-Qaeda Osama Bin Laden, but also tens of thousands of other human beings. British Prime Minister Tony Blair joined in this offence and adopted the narrative of the moral, defensive war that would serve the greater good, as he and President Bush prepared to invade Iraq. But the 2003 invasion and occupation of Iraq did not serve the greater good. What did it do and what role did drones play? That is the topic of this chapter.

The 2003 Invasion and Occupation of Iraq

Having 'won' the Cold War, having made parts of the world safe and ready for democracy, the call to the United States to spread American values, while simultaneously serving American interests, compelled a new crusade.[4] In this crusade the United States was to strive for universal freedom, but also for universal dominion, a world order described by Francis Fukuyama in 'The End of History', which proclaimed that the Hegelian end of history was occurring in a world at such a stage of sociocultural evolution that perfect civil society was possible. Liberal democracy would be the final form of human government, bringing with it the end of war, the West's ideological debates and a common marketisation of the world.[5] America's vision now was to steer the world away from a multipolar system to a unipolar world, where a strong West would dominate, guided by the greatness and power of the United States; a 'super-sovereign West economically, culturally, and politically hegemonic in the world'.[6]

Attacks would have to be on terrorist-harbouring countries, not terrorist networks. It was important to capture Bin Laden and to defeat Al Qaeda, but the overriding aim of the War on Terror was to change regimes, as neither universal freedom nor universal dominion could be achieved without regime changes. When Colin Powell demanded that the Taliban give up bin Laden and Al-Qaeda, journalist and political columnist Charles Krauthammer replied that the Taliban had to pay even if it delivered both: 'If the administration goes wobbly on the Taliban, it might as well give up the war on terrorism before it starts … The take-home lesson must be: Harbor terrorists – and your regime dies.' Krauthammer wanted the United States to overthrow Syria's Assad regime, low-hanging fruit that harboured terrorist groups. After Syria, the United States would have to remove the regimes in Iran and Iraq. 'The war on terrorism will conclude in Baghdad', he predicted. 'If this president wants victory in the war he has declared, he will have to achieve it on the very spot where his own father, 10 years ago, let victory slip away.'[7]

What the War and the End of History required was nothing less than complete regional transformation. Of the twenty-two Arab states, none was a democracy, while nearly all were hostile to Israel. The war against terrorism was a war against Arab-grown radical Islamism, and the key to changing the Middle East was to overthrow the Baathist regime in Iraq: 'A de-Saddamized Iraq with a decent government could revolutionize the region.'[8] On overthrowing Saddam Hussein, the United States would have a base for the outward projection of American power and the dissemination of democratic and modernising ideas. Global subordination to American power was to be America's imperial burden.

This was imperial realism: '9/11 was a wake-up call for the United States to unambiguously embrace its imperial responsibilities.'[9] The call was to consolidate American power in the Middle East by changing political culture in the region. The War on Terror has been described as a neo-imperial programme, designed to further American political interests. By invading the Middle East, the United States would gain physical and strategic control over energy supplies, as it transformed regimes, brought peace, liberty and democracy. It would be the perfect marriage of might and right.

America's hegemonic position rests on its ability to control the sources and transport routes for crucial energy and other material supplies needed by leading industrial states. For this it needs strategic positioning to achieve and maintain sea, air and land dominance, especially in the Middle East, the eastern Mediterranean, the Atlantic, Pacific and Indian Oceans. The 9/11 attacks gave the United States the opportunity to try to monopolise energy resources, to increase its grip on the Middle East and to demonstrate that it had an important role to play in the post-Cold War world, one that combined military force and political and economic control. The 9/11 attacks were an opportunity for the United States to declare its (Western) values universal and to start to enforce them, which was consistent with its neo-imperial ideology. Thus, it was going to dominate by making its ideas and values the dominant ideas and values:

> This is an era of 'tremendous opportunity' for America, inasmuch as America can present its national values as universal ones and impose them on the globe, also by means of violence. Again, this is not a new ingredient of imperialism: It is an ages-old and enduring ideological principle that conditions the political attempt by a group of people to become dominant either nationally or internationally. To dominate others, one's ideas must become dominant over other people's ideas.[10]

As a global hegemon, the United States would have the power to define what was true and right, as well as gain control of resources and the global economy, by applying military force, with as much loss of life as was necessary. The War on Terror may be infinite; it would last as long as necessary for the United States to achieve full dominance and imposition of its values of liberal democracy and free-market economy on the globe. It has now become almost clichéd to say that the War on Terror was declared by the United States so that it could gain access to Iraq's oil. Yet plans to take control of the oil were made a few months before the events of 11 September 2001, by Dick Cheney's Energy Task Force, which discussed the 'capture' of Iraqi oil from February 2001: 'By May 2001 it had already set out, urgently and in some detail, plans for taking control over Iraqi oil.'[11]

Iraq was a prize bigger than Taliban-controlled Afghanistan. Due to Iraq's strategic position in the Persian Gulf and the Middle East, Saddam Hussein represented a bigger challenge to the United States and its interests and principles than the weak, isolated Taliban. The door to a worldwide US war against terrorism opened with 9/11: 'The war on anti-American terrorism must target Hezbollah, the terrorist group backed by Iran and Syria, as well as the Taliban. And it must include a determined effort to remove Saddam Hussein from power.'[12] Regime change in Afghanistan and Iraq was by no means sufficient, but it was a necessary condition, if political, economic and cultural transformation/domination were to begin.

George W. Bush had arrived at the White House with a pledge to put $20 billion of the defence budget towards technology research, claiming that the United States must hold on to its technological edge in an unstable world and create the military of the next century. In December 2001, after the War on Terror had been declared, Bush, in a speech at the Citadel, declared:

> our special forces have the technology to call in precision air strikes – along with the flexibility to direct those strikes from horseback, in the first cavalry charge of the 21st century. This combination – real-time intelligence, local allied forces, special forces, and precision air power – has really never been used before. The conflict in Afghanistan has taught us more about the future of our military than a decade of blue ribbon panels and think-tank symposiums. The Predator is a good example. This unmanned aerial vehicle is able to circle over enemy forces, gather intelligence, transmit information instantly back to commanders, then fire on targets with extreme accuracy ... We're entering an era in which unmanned vehicles of all kinds will take on greater importance – in space, on land, in the air, and at sea. Precision-guided munitions also offer great promise. In the Gulf War, these weapons were the exception – while in Afghanistan, they have been the majority of the munitions we have used. We're striking with greater effectiveness, at greater range, with fewer civilian casualties. More and more, our weapons can hit moving targets. When all of our military can continuously locate and track moving targets – with surveillance from air and space – warfare will be truly revolutionized.[13]

Afghanistan had already been invaded, by the time President Bush made that speech. According to the Project on Defense Alternatives, between 7 October 2001 and 1 January 2002, 1,000–1,300 civilians were directly killed by the US-led aerial bombing campaign[14] and by mid-January 2002, at least 3,200 more Afghans had died of starvation, exposure, associated illnesses, or injury sustained while in flight from war zones.[15]

In this War on Terror context of morality, truth and noble missions, on one hand, and interests, dominance and the deaths of thousands of innocents, on the other, the invasion of Iraq – Shock and Awe – officially began on 19 March 2003. They called it Operation Iraqi Freedom. By the time the invasion was completed on 1 May, some 7,500 Iraqi civilians had been killed in the airstrikes.[16] However, the bombings had started earlier in the year. Iraq Body Count documented the killing of five Iraqi civilians by precision guided weapons on 1 and 6 January, and 10 February 2003.

Table 4.1: Iraq Body Count Incident x001[17]

Incident	x001
Type	'precision guided weapons'
Deaths recorded	1
Targeted or hit	'military air defence radar'
Place	Al Qurnah, 210 kilometres south-east of Baghdad
Date and time	1 January 2003, 6:25 a.m.+

Table 4.2: Iraq Body Count Incident x002[18]

Incident	x002
Type	'precision guided weapons'
Deaths recorded	2
Targeted or hit	'mobile radar equipment'
Place	near Al Amarah, 265 kilometres south-east of Baghdad
Date and time	6 January 2003, p.m.

Table 4.3: Iraq Body Count Incident x003[19]

Incident	x003
Type	'precision guided weapons'
Deaths recorded	2
Targeted or hit	'mobile surface-to-air missile system'
Place	near Basra
Date	10 February 2003

The pre-invasion killings may have been few, but the deaths resulting from Shock and Awe were more like the incident below.

Table 4.4: Iraq Body Count Incident x038[20]

Incident	x038
Type	air raids
Deaths recorded	226–40
Place	Nassiriya
Date	20 March 2003 to 3 April 2003

Identifying details		
x038-uw359	Brother of Haider Mohammed	Male
x038-dn276	Brother of Haider Mohammed	Male

Other information		
Number killed	**Age**	**Sex**
1	11	Male
1	3	Female
55	Child	*Unrecorded*
37	Adult	Female

Among the early victims of the invasion were two children, sister and brother Lina and Mohammed Wail Mwafaq Mosa Tabra, killed on the night of 31 March 2003, as they sheltered in a farm in Tikrit. They were among nineteen people killed from an airstrike that took the lives of five Iraqi children.

Table 4.5: Iraq Body Count Incident a6384[21]

Incident	a6384
Type	Coalition air strike involving two missiles
Deaths recorded	19
Targeted or hit	Group of families sheltering in a farm compound
Place	Al-Ishaqi, south of Tikrit
Date and time	31 March 2003 to 1 April 2003, around midnight

(continued)

Identifying details

IBC page	Identifying details	Age	Sex
a6384-zv3670	Lina Wail Mwafaq Mosa Tabra	14	Female
a6384-sz3485	Mohammed Wail Mwafaq Mosa Tabra	9–10	Male
a6384-ev3496	Sameera Rasheed Faraj	40–50	Female
a6384-dv3539	Ban Rasheed Faraj	37–38	Female
a6384-ue3666	Aseela Sami Mohammed	5–8	Female
a6384-dz3481	Mother of Mayyasa and Mohammed Bashar (pregnant)	20–40	Female
a6384-zs3492	Mayyasa Bashar	10	Female
a6384-xe3543	Mohammed Bashar	2	Male
a6384-za3678	Hajji Abbas	65–75	Male
a6384-xa3477	Wife of Hajji Abbas	60–75	Female
a6384-ea3488	Khalid Abbas	30–40	Male
a6384-dx3547	Relative of Hajji Abbas Family	*Unrecorded*	Female

'Uncomplicated' and 'a quiet girl', was how her father described Lina eighteen years later. 'As a teenager she didn't share her interests with us, but after her death, her teenage cousins told us that she liked to keep a diary and write poems' (as told by her father Wail Mwafaq Mosa Tabra to Iraq Body Count, 4 July 2021).

The invasion of Iraq was planned in accordance with the concept of rapid dominance – rapidity and timeliness in application, operational brilliance in execution – and coordinated by an HVT cell in the Pentagon. In attempting to hit Saddam Hussein through an HVT, due to the kill-chain time (the time between getting the intelligence and the missile impacting) being too long (around forty-five minutes) many civilians were hit instead. In *Kill Chain*, Cockburn quotes a Pentagon analyst, who recalls 'The shortest kill chain we managed in the 2003 war was forty-five minutes. That was the strike on the al-Saath restaurant in Baghdad. We thought Saddam was there. He wasn't, but we did kill a bunch of civilians.'[22] Drones were going to shrink the kill chain to zero, by waiting for the target to appear and then launching a missile, but without having the time to truly assess the collateral damage, even if the right target was hit. During and following the invasion, precision strikes targeted purported lairs of various Iraqi Commanders, as well as Saddam Hussein, without success. All fifty high-value individuals survived. Not so lucky, according to former Defense Intelligence Agency analyst Marc Garlasco, were the 'couple of hundred civilians, at least' who were killed in the strikes.[23] Lina, Mohammed and the other seventeen civilians killed on 31 March may have been among those civilians – collateral damage in a hegemon's war.

Regulations stipulated that civilians could be killed, but not too many (not more than thirty), not without clearance from higher authority. Garlasco explains, 'If you're gonna kill up to twenty-nine people in a strike against Saddam Hussein, that's not a problem. But once you hit that number thirty, we actually had to go to either President Bush, or Secretary of Defense Rumsfeld.'[24] The clearance needed to risk the lives of over thirty civilians was frequently requested and was never refused.

An Iraq Body Count dossier revealed the extent of the killings between 2003 and 2005. During the invasion and in the two years that followed, 24,865 civilians were reported killed – almost half in the capital Baghdad. The US-UK coalition forces killed 37 per cent of all civilian victims in the first two years. Anti-occupation forces and insurgents killed 9 per cent, post-invasion criminal violence accounted for 36 per cent of all deaths, and the remainder were killed by unknown agents. At least a further 42,500 civilians were reported wounded.[25] In the years that followed, tens of thousands were added to the civilian death toll and by the end of 2013 Iraq Body Count had documented 134,571 civilian deaths by the US-UK coalition, Al Qaeda in Iraq, Mahdi Army, Iraqi government forces and a variety of militia and insurgent groups, through air strikes, raids, suicide bombings, car bombs, improvised explosive devices, shootings, abductions and executions. See Table 4.6.

Meanwhile, Barack Obama won the 2008 Presidential election. To distinguish himself from self-declared 'War President' Bush, he had run a quasi-anti-war campaign for the White House; however, he embraced and increased the assassination programme. Under Obama, according to intelligence analyst Robert Gates who served as Secretary of Defense 2006–11, factories were working day and night to produce the weapons to fight terrorists. 'From now,' he declared, 'the watchword is: drones, baby, drones!' as he accepted an award for exemplary service to the nation and the CIA at the Richard M. Helms Award dinner on 30 March 2011.

Table 4.6: Monthly civilian deaths 2003–13[26]

	Jan	Feb	Mar	Apr	May	Jun	Jul	Aug	Sep	Oct	Nov	Dec	
2003	3	2	3,986	3,448	545	597	646	833	566	515	487	524	12,152
2004	610	663	1,004	1,303	655	910	834	878	1,042	1,033	1,676	1,129	11,737
2005	1,222	1,297	905	1,145	1,396	1,347	1,536	2,352	1,444	1,311	1,487	1,141	16,583
2006	1,546	1,579	1,957	1,805	2,279	2,594	3,298	2,865	2,567	3,041	3,095	2,900	29,526
2007	3,035	2,680	2,728	2,573	2,854	2,219	2,702	2,483	1,391	1,326	1,124	997	26,112
2008	861	1,093	1,669	1,317	915	755	640	704	612	594	540	586	10,286
2009	372	409	438	590	428	564	431	653	352	441	226	478	5,382
2010	267	305	336	385	387	385	488	520	254	315	307	218	4,167
2011	389	254	311	289	381	386	308	401	397	366	288	392	4,162
2012	531	356	377	392	304	529	469	422	400	290	253	299	4,622
2013	357	360	403	545	888	659	1,145	1,013	1,306	1,180	870	1,126	9,852

Just days after Obama's inauguration, in January 2009, two drone strikes that he authorised killed up to twenty-five people, including as many as twenty civilians in Waziristan, Pakistan. Neither strike hits its HVT; the first strike hit the home of eighteen-year-old student Faheem Qureshi, killing his cousins and friends and blinding him in one eye, while the second took the life of local elder Malik Gukistan Khan, who was a member of a peace committee. Four members of his family, his nephew and three sons (the youngest just three years old) died with him.[27] The strikes continued, then doubled and redoubled, until life in Waziristan became 'hell on earth'.[28] The remote-killing campaigns devastated the life of the society and traumatised communities, as if they had been subjected to Second World War-style bombings, but in ways that would be invisible to distant spectators peering at their Predator feeds. The double-tap tactic of reserving a second missile for anyone coming to rescue the wounded or retrieve the dead bodies, 'put a crimp at the generosity of ordinary citizens, not to mention the Red Cross, which ordered its people to stay away from a house or car hit by drones for at least *six hours*'.[29] Inevitably, collateral victims accumulated: vehicle drivers and passengers, children on their way home, men and women in open markets, some burnt so badly their bodies were unrecognisable. Over fifteen days in the summer of 2013, US drone strikes hit Yemen nine times, killing forty-nine people, three of whom were children, while in December of the same year four missiles hit a wedding convoy killing twelve men.[30]

In Iraq, as British and American troops were withdrawing 2009–11, killing levels dropped to 4,000–5,000 civilians a year, the lowest since Iraq had been invaded. Yet as al-Maliki's democratically elected government of Iraq escalated its attacks against its citizens killing more than 1,300 of them, the death toll doubled in 2013, as nearly 10,000 people lost their lives. But the worst was about come, in the form of the Islamic State of Iraq and Syria and the battles that followed. See Table 4.7.

It could be argued that June 2014 was the bloodiest since the invasion: 4,088 civilians were killed in one month, according to Iraq Body Count records (in March 2003, Operation Iraqi Freedom had killed 3,986 in twelve days). While ISIS is still in Iraq, it was in the 2014–17 period that it did most of its killing, as it ruled over its Caliphate. ISIS was and remains unparalleled in its brutality. In Iraq alone, it has killed more than 50,000 civilians, including

Table 4.7: The ISIS years (monthly civilian deaths 2014–17)[31]

2014	1,097	972	1,029	1,037	1,100	4,088	1,580	3,340	1,474	1,738	1,436	1,327	20,218
2015	1,490	1,625	1,105	2,013	1,295	1,355	1,845	1,991	1,445	1,297	1,021	1,096	17,578
2016	1,374	1,258	1,459	1,192	1,276	1,405	1,280	1,375	935	1,970	1,738	1,131	16,393
2017	1,119	982	1,918	1,816	1,871	1,858	1,498	597	490	397	346	291	13,183

entire families. The violence it brought to the country in 2014 was hardly new, but one type of killing became its 'signature': executions. In the database of Iraq Body Count many such incidents have been documented, such as the six women and girls executed on 22 October 2016.

Table 4.8: Iraq Body Count Incident a5888[32]

Incident	a5888
Type	gunfire, executed
Deaths recorded	6
Targeted or hit	women and children executed for falling behind during a forced relocation, because one of the girls had a disability
Place	al-Rufaila village, Tulul Nasir, south of Mosul
Date	22 October 2016

Personal details

a5888-xw3516	disabled girl	Child	Female

Demographic information

Number killed	Age	Sex
3	Adult	Female
2	Child	Female

On some days the executions reached triple figures, such as the 20 to 21 October 2016 killings that included sixty-one children – all were buried in mass graves. See Table 4.9.

It was easy to become an ISIS victim; the group executed those accused of a number of infractions: trying to flee, being homosexual, not being sufficiently covered in public (women and girls), shaving or cutting one's hair (men), listening to music, using a mobile phone, playing or watching football, committing adultery, collaborating with enemies of ISIS (such as Iraqi security forces), and doctors were executed for refusing to treat injured ISIS fighters. Methods of execution, depending on the infraction, included

Table 4.9: Iraq Body Count Incident a5885[33]

Incident	a5885
Type	gunfire, executed
Deaths recorded	215–81
Targeted or hit	men and children from areas and villages of northwest of Mosul, including al-Zawiyah village, bodies of victims were then transferred to mass graves by a bulldozer
Place	Agriculture College, north Mosul
Date	20 October 2016 to 21 October 2016

Demographic information		
Number killed	Age	Sex
61	Child	Male
220	Adult	Male

being shot, electrocuted, thrown off a building, drowned, set on fire and having one's head crushed between two rocks.

The British Parliament began to debate further intervention in Syria in September 2014, the resuming of air strikes, in the form of precision strikes, but in Iraq had already commenced. Iraqi civilians were no longer getting killed by terrorists alone, after August 2014. The airstrikes meant for ISIS took the lives of thousands of innocents, such as Mohannad Rezzo, a university professor, and his seventeen-year-old son, Najeeb; his sister-in-law Miyada and her twenty-one-year-old daughter, Tuka. The four family members were killed when a drone strike flattened their home as they slept on 21 September 2015. Here is an incident of a drone hitting a medical centre on 18 October 2016. See Table 4.10.

As people were blown up by car bombs and suicide bombers, as they were shelled, shot and executed, they were also killed by those who claimed to be there to protect them. See Table 4.11.

An estimated 13,000 Iraqi civilians lost their lives in drone strikes between 2014–17.[34]

Did the strikes at least end or reduce the overall killings of civilians? The result of the renewed intervention was not the creation

Table 4.10: Iraq Body Count Incident a5878a[35]

Incident	a5878a
Type	air attacks
Deaths recorded	3
Targeted or hit	Hammam Al-Alil medical centre hit, casualties include children and medical centre staff
Place	Hammam Al-Alil, south of Mosul
Date and time	18 October 2016, 09:30 a.m.

Demographic information

Number killed	Age
3	Child

Table 4.11: Iraq Body Count Incident a6317[36]

Incident	a6317
Type	air attacks
Deaths recorded	30–49
Targeted or hit	residential areas of displaced people hit, casualties include women and children
Place	Dawrat Qassim al-Khayat and Al-Shifa areas, west Mosul
Date and time	19 February 2017, a.m.

Demographic information

Number killed	Age	Sex
7	Child	*Unrecorded*
3	Adult	Male

of a peaceful state, the drone strikes did not stop or decrease the violent deaths of the unarmed population, they did not abate sectarian conflict, and they did not alleviate suffering. After the coalition renewed its air strikes in Iraq, the country witnessed its largest bombing in years: 120 civilians were blown up by a suicide bomber on 17 July 2015, in Khan Bani Saad. Additionally, Iraqi government forces killed another 1,532 civilians in 2015, in the fight against ISIS. On a more positive note, while more than 4,000 US military personnel had been killed in the invasion and occupation of Iraq 2003–11, only a few were killed in the resumption of air strikes in the battle against ISIS and those were all on the ground: Petty Officer 1st Class Charles Keating (aged thirty-one) was killed in fighting near Erbil; Marine Staff Sergeant Louis Cardin (aged twenty-seven) was killed when ISIS militants fired rockets at a coalition base in Makhmour; Army Master Sergeant Joshua Wheeler (aged thirty-nine) was killed during a raid on an ISIS compound in Hawijah.[37]

The Vicarious Warfare of a Superpower

America's hybrid, persistent, remote and evasive mode of fighting

Following the invasion of Iraq, the United States has fought its hybrid wars through a combination of drone strikes, special forces, intelligence operatives, private contractors and military-to-military training teams in many countries. Kiras refers to this policy as seeking shortcuts to victory: 'new means, including special operations and/or emerging technologies, purportedly allow users to avoid the difficulties posed by the nature of strategy'.[38] In the aftermath of the costly wars in Afghanistan and Iraq, there was a shift towards tech-enabled economisation, privatisation and distance, as state power met imperial ambition, seeking ways to confront strategic challenges. Vicarious warfare is associated with powerful states, especially those that enjoy forms of imperial dominion: great power brings incentives and requirements central to a hegemonic status, that are pursued predominantly in a forceful manner.

> Such crusading liberalism continues to inspire US policymakers in reversing all manner of global evils – whether authoritarianism, oppression, ethnic cleansing or Islamic terror – and promoting ideals of neoliberal economics, republican democracy and freedom. It is an agenda that has manifested itself in military terms most predominantly in modern times through what Freedman has termed 'offensive liberal wars'.[39]

In particular, where force is deployed in pursuit of political and economic interests, the means adopted do not entail national sacrifice, but rather vicarious approaches that normalise war. A more general American faith in the efficacy of force and its belief in its unconquerable superiority means that the military is regarded as a tool that is perfect for fixing things and solving problems affordably. The moral conviction that accompanies the use of force 'encourages the conceit that America is only capable of wielding a righteous sword against those who must, by default, be evil'.[40] Vicarious, virtual, moral warfare (so the narrative goes) is favoured in the pursuit of primacy and the preservation of *Pax Americana*, in a culture of patriotism, risk-aversion and bloodless victories. The representation of smart bombs and surgical strikes sustains and perpetuates the myths of discriminate weapons and moral righteousness, and encourages popular triumphalism.[41] Instead of revulsion regarding war and all its horrors, high-tech, 'humane' campaigns become an acceptable and respectable spectacle. Hanson observes that in America there is great popular enthusiasm for all things military, which may express 'an innate need to experience violence, if only vicariously'.[42] The new American militarism can be seen in the proliferation of shooter video games that propagate sanitised images of warzones and battlefields, in parallel with remote warfare, leaving the American public largely unaffected by America's wars. Where the public is affected is in a romanticised relish in the military campaigns embarked on in the nation's name. Moral disengagement, unaccountability and ignorance regarding the consequences of the nation's unceasing interventionism, combined with the low cost in blood and money, in the words of Holmes, 'appeases an indifferent electorate'.[43] The vicarious warfare model with its minimalist, low-exposure operations 'allows the American people to place themselves at a remove from the difficult decisions that, in a democracy, should accompany the use of force'.[44] By distancing society from the country's wars, any moral responsibility for the consequences of those wars is diluted, there are no political drawbacks, and the vicarious war can continue on an indefinite basis.

By the end of 2011, US troops had been withdrawn from and Iraq and Afghanistan, but thousands of defence contractors remained in both countries, responsibility for major combat operations was passed on to indigenous security forces, while the drone programme was bolstered and expanded in campaigns in Libya and Tunisia, in the context of the Arab Spring – in the latter cases (and in contrast to Iraq and Afghanistan), helping anti-regime rebels. As the United States fought ISIS in Iraq, in 2017 President Trump expanded the scope of US military efforts to Somalia, to fight al-Shabaab using special forces and air strikes. US Africa Command (AFRICOM) conducted thousands of air strikes, but only admitted to killing civilians in a single strike that took place in April 2018, under pressure from Amnesty International. Deprose Muchena, Amnesty International's Director for East and Southern Africa, stated, 'We've documented case after case in the USA's escalating air war on Somalia, where the AFRICOM thinks it can simply smear its civilian victims as "terrorists," no questions asked. This is unconscionable.'[45] In these two cases from February 2020, the precision of the so-called 'humane attacks' can be seen.

- Case 1: At around 8:00 p.m. on 2 February 2020, a family of five was having dinner in the city of Jilib, when an air-dropped weapon – likely a US GBU-69/B Small Glide Munition with a 16-kilogramme warhead – struck their home. Nurto Kusow Omar Abukar (aged eighteen) was struck in the head by a heavy metal fragment from the munition and killed instantly. The strike also injured her younger sisters, Fatuma (aged twelve) and Adey (aged seven), and their grandmother, Khadija Mohamed Gedow (aged seventy).
- Case 2: On 24 February 2020 a Hellfire missile hit the Masalanja farm near the village of Kumbareere, 10 kilometres north of Jilib, killing Mohamud Salad Mohamud (aged fifty-three). He was a banana farmer and Jilib office manager for Hormuud Telecom, and he left behind a wife and eight children.[46]

In the past fifteen years, the United States has fought its wars completely remotely through embracing drones as 'magic bullets', using proxy forces and adopting indirect approaches, in what Waldman calls '3-D' warfare, where remoteness is achieved through danger-proofing, delegation and darkness. An extreme form of danger-proofing, the use of airpower and long-range weapons systems such as cruise missiles and drones, allows war to be conducted at a safe distance from direct physical threat, in accordance with the US military's force protection fetish:

Airpower as a seductive shortcut to victory ... Today remote control methods promise a form of proximity without risks. Some enthusiasts proclaim the ability to effect 'air occupations', while others tout the special benefits of America's 'asymmetric advantage' in airpower that can 'present a truly show-stopping impediment to the nefarious schemes of her enemies' with 'impunity and little risk to Americans'.[47]

Cruise missiles and drones, however, can only coerce and kill. An overreliance on this kind of warfare results in an inability to affect political dynamics on the ground. In addition, the grievance and terror that this air occupation instils in those who have to live with the surveillance and perpetual mortal threat from the sky, often provoke hatred, opposition and insurgency, rather than compliance or surrender.

Vicarious warfare also involves shifting or externalising the burden of risk and responsibility onto others, by contracting security tasks to proxy actors. In Iraq, the United States transfers the risks of combat to local allies, private military and security companies, local security forces or irregular militias. This not only averts risk, it also hides the true costs of war. In 'Expendable soldiers', Porch argues that using proxies lowers the political and financial costs of intervention, by desensitising the population to the human overheads of foreign adventures.[48] Pro-Western – or at least anti-jihadist – locals are trained and equipped to do the fighting, the killing and the dying. Those men, expendable as they are, are presented as loyal partners who will follow American orders and share American values and objectives. The use of private military and security companies enables the invisibility (to domestic audiences) of operations and casualties; thus, shielding policymakers from unwelcome press. The Pentagon rarely provides data on contracted operations. When popular support for war depends on casualty figures and troop numbers, opaqueness in reporting and lack of awareness regarding the human cost of America's privatised wars both help gain that support and subvert any real democratic debate. The lack of transparency permits the United States to conduct wars that the public might oppose if they knew the full scale of the commitment and the casualties.

While large Western firms are awarded big security contracts, most of the personnel are third-country nationals or host-nation citizens; jobs are subcontracted to local security companies through opaque outsourcing chains. In Afghanistan the United States hired warlords who had links with insurgents, with the Taliban, or who were implicated in murder and other violent crimes. In Iraq the risky tasks of defending, of killing and of managing local security in terrorist-riddled areas were passed on to the Awakening Councils, the Popular Mobilisation and government forces. Building the indigenous military capacity of foreign state forces to defend themselves or confront threats in line with US interests is accomplished through training, arming and advising. The United States is not alone in this practice of remote warfare: in the twenty-first century, the United Kingdom has built many such 'partnerships'. In 'Avoiding civilian harm in partnered military operations', Day, Ledwidge, Casey-Maslen and Goodwin-Hudson reveal that partnered military operations have become the predominant form of UK military engagement, 'seeing the UK training, arming and fighting alongside allied forces, other state militaries and armed opposition groups across the world'.[49] UK partnered military operations involve activities such as training, advising, assisting, accompanying, providing kinetic, logistics and intelligence support, and partnering detention operations. Working with and through local and regional forces who do the bulk of the fighting, while the United Kingdom and other Western partners, such as the United States, provide support, seems rational and low risk.

Preparing and partnering with local forces for aggressive kinetic operations is more widespread than the public knows. Just taking the United Kingdom as an example, a country much less powerful than the global hegemon, the Ministry of Defence disclosed in response to parliamentary questions on 13 July 2020 and 25 March 2021 that 145 countries have received training from British armed forces, including Afghanistan, Iraq, Armenia, Azerbaijan, China, Israel, Nigeria, Pakistan, Palestinian Authority, Saudi Arabia, Serbia, Somalia, Sudan and South Sudan. What is noteworthy is that some of those the United Kingdom has partnered with are states that are fighting each other and/or have very poor human rights records. That the United States and the United Kingdom have often resorted to employing irregular surrogate forces such as rebels, militias, paramilitary groups and warlords, and to partnering with militaries of states with poor human rights records, displays a concern with short-term gains and a lack of concern for serious short- and long-term harm:

Undertaken in the midst of punishing campaigns, the emphasis has been on rapidly fielding large conventional forces in pursuit of immediate security rather than ensuring quality or long-term sustainability. The result has been the creation of fragile forces plagued by corruption, political divisions, enemy infiltration and operational deficiencies.[50]

In some cases, partnerships have proven counterproductive. Knowledge of such support can undermine the regular or irregular groups that they are seeking to empower: locals receiving support are regarded as collaborators and puppets – and that includes state security forces, such as those of Iraq post-2006. Militia actions have also contributed to instability, insecurity, oppression and casualties. Empowering irregular groups may achieve immediate military objectives, but can create political problems, such as 'destabilizing rebel fragmentation, proxies making unpalatable demands on attaining power or their pursuit of illiberal exclusionary agendas'.[51] It can also create ethical problems, such as abetting unlawful and immoral conduct against civilians, assisting violations of international humanitarian law and partnering in operations that pose great risk of death and/or injury to civilians.

Covert interventions, through direct force, by proxy or in cyberspace, undermine fundamental international laws and norms that Western states, chief among them the United States, claim to uphold. The darkness of offensive cyber warfare is another example of how operations take place in the shadows and outside oversight and accountability. This blurring of boundaries and merging is, in operational terms, what has been called the 'global shadow war', where CIA teams, supported by special forces, conduct lethal drone strikes as part of an ongoing campaign of targeted killings, mainly outside active war zones. The shadow warriors are flexible, rapidly deployable and highly capable military instruments, but they also promise major results that can be obtained for minimal outlay,[52] a tempting 'easy button' for intervention.[53] Such secretive, aggressive missions have substituted any strategy that might take into account regional and local requirements and priorities. Much like covert operations and drone warfare, weaponised malware provides policymakers with discrete, surgical weapons that project national power with secrecy and deniability.

According to revelations and leaks by Edward Snowden 2013–14, the US intelligence services conducted more than 200 offensive cyber operations in 2011. More recently, the Pentagon's Cyber Command has adopted an even more aggressive cyber strategy, with a growing arsenal of offensive cyber weapons that it has deployed against terrorist websites and North Korea's Musudan ballistic missile systems. It most famously conducted a fully automated covert operation against the Iranian nuclear centrifuges in the form of the Stuxnet virus. Stuxnet was created by the intelligence agencies of the United States and Israel to destroy the centrifuges that Iran was using to enrich uranium as part of its nuclear programme. In the United States, development started under George W. Bush and continued under Obama. The classified programme to develop the virus was named Operation Olympic Games. The Bush and Obama administrations believed that if Iran were on the verge of developing atomic weapons, Israel would launch airstrikes against Iranian nuclear facilities in a move that could have set off a regional war. Operation Olympic Games was seen as a non-violent alternative.[54]

Cyber weapons are among those ethical, humane, non-violent, low-risk weapons (or so it is claimed) that seek to confront adversaries and threats in the dark. Remoteness and darkness have characterised America's warfare. The distance and secrecy, as in the case of cyberattacks, and lack of disclosure around cyber policy, drone and covert operations has meant that the efficacy and the ethics surrounding the use of such methods has evaded robust democratic debate, scrutiny and accountability. Operational darkness has gone hand in hand with political darkness and moral darkness. It has created deepening mistrust and led to serious escalation resulting in thousands of victims, directly through bombings and abuses, or indirectly through the psychological trauma of protracted conflict. In Iraq, the conflict continues to this day, in a war that most of the world regards from a safe distance, geographically and by the passage of time, fading from sight and memory. Over twenty years, barely a day has passed without reports of civilians being shot or blown up. Personal loss, the desire for retribution and numerous grievances mean that multiple killings continue, unlawful killings and executions, as well as those sanctioned and carried out by the state. Neither kind interrupts the cycle of violence, instead drawing new participants into it. Other deeply felt grievances relate to ethnic or religious persecution; the toxic residues of military occupation and foreign domination; economic disparity; political and judicial corruption.[55] An entire generation of Iraqi children has grown up in a country riven by violence, fear, displacement and poverty. In 2023, five years after the battle for Mosul ended, bodies are still being pulled from the rubble of their homes.

Meanwhile America's status as the greatest global military power has come to signify an affirmation of US exceptionalism:

> At least as measured by our capacity to employ violence, we are indeed Number One. The providential judgment seems indisputable: the nation charged with the responsibility for guiding history to its predetermined destination has been endowed with the raw power needed to do just that.[56]

Within its military supremacy, the United States regards its armed forces as the highest expression of state power and virtue, made possible by its military technology, relying on force to spread the American Way of Life, to protect American interests and to impose its will 'at gunpoint if need be'.[57] In the realist tradition, preparing for war and waging war have become its normal state.

Empire, Militarism and American Virtue

Reinventing war and peace

> The only accepted 'plan' for peace is the loaded pistol.[58]

Following the end of the Cold War, the United States, empowered by 'the end of history' and its new position as the global leadership, scattered its soldiers across Europe, the Far East, South America and the Persian Gulf, to assume its hegemonic responsibilities. The Bush doctrine of preventive war became the mantra of the new American militarism:

> As a statement of intent, the doctrine is unambiguous: in an age when deterrence 'means nothing' and containment 'is not possible', the United States will exercise the prerogative of striking first. 'In the world we have entered,' George W. Bush has declared, 'the only path to safety is the path of action. And this nation will act.'[59]

The aim of preventive war, as Iraq showed, was to kill – and quickly. Force would be employed routinely and proactively to defeat rivals and aggressors, to pursue interest, to replace hostile regimes and to spread Western political and moral values. The American Empire grounded on military pre-eminence can strike and kill at will, doing so not only to increase its own security, but also on behalf of the oppressed, demonstrating its mastery and its humanity, yet not taking into account the costs incurred along the way.

When Obama campaigned for presidency, his message was one of hope, of change and for a new chapter, particularly regarding US military policy, following the departure of War President Bush. Once Obama took office, after winning the 2008 election, he was expected to reverse course. In keeping with his promise for peace, he was awarded the Nobel Peace Prize in 2009. Under his leadership, the Pentagon was going to abandon 'outdated Cold War-era systems so that we can invest in the capabilities that we need for the future'. A 'leaner' military establishment was promised, but US forces would remain capable of handling 'the full range of contingencies and threats'.[60] He declared the United States 'the greatest force for freedom and security that the world has ever known'. It soon became clear that his administration's new strategy was not to end the book of war, but to start a new war chapter. Obama's approach to national security differed from Bush's, but it preserved more than it changed. 'Has President Obama adopted George W. Bush's "policeman of the world" approach to the fight against terrorism?' *Washington Post*'s Walter Pincus asked.[61] According to his national security strategy, 'Military involvement may be necessary to stop a bloody conflict, but peace and stability will last only if follow-on efforts to restore order and rebuild are successful' and the United States would consider 'directly striking the most dangerous groups and individuals when necessary'. Pincus went on to ask who those dangerous groups and individuals were: 'Who puts them on the list, and what are the criteria? And who makes the decision that direct U.S. strikes are needed, and on what basis?' The world would become a battlefield, where the United States could strike anyone in 'self-defence'.

A new, virtuous America was reinventing war and peace, by substituting bloody, victorious short wars for indecisive long wars that resembled an uneasy peace. There was no reluctance to use force, only a reluctance to wage total war and risk American lives. The lives of the 'dangerous' individuals would not be spared. Those smaller wars would be a lot more targeted and far more acceptable to the public. The Obama administration set out to establish 'a constellation of secret drone bases in and around the Arabian Peninsula and the Horn of Africa – new platforms from which to conduct attacks against Islamic radicals wherever they might be found'.[62] Yet the stepping up of the drone killing programme to carry out executions of people presumed guilty, came as a shock to those anti-war activists that had supported President Obama. Eventually, however, the public came to accept both the targeted assassinations programme and the 'collateral damage' when it went wrong, as they came to accept that 'a small number of human beings possess the right to decide who must die and what should be an acceptable price to pay in *other* people's lives in the quest for a sought-after goal'.[63] Rather than pursue criminals within the framework of international laws, American presidents decided to capture, torture and imprison without charge, to embark on killing campaigns worldwide and to label evil those who are against its security and other interests, thus justifying their execution.

In *The Sorrows of Empire* Chalmers Johnson refers to the United States as 'a military juggernaut intent on world domination'.[64] In this quest, it has used sanitised expressions such as 'preventive war', 'collateral damage' and 'regime change', to describe and justify its aggression: invasions, occupations, violations of sovereignty and crimes against civilians. This 'militarised empire' is 'a network of economic and political interests tied in a thousand different ways to American corporations, universities, and communities' but kept separate from everyday life.[65] The United States has been using imperial methods, such as the establishment of military garrisons around the globe, the granting of subsidies to client governments, the application of economic sanctions, as well as military force against recalcitrant states. Moreover, 'its advocates never question the virtues of empire ... and they do not for a moment doubt that it is in the best interests of those over whom it rules'.[66] Johnson calls this an empire of bases, consisting of surrogate soldiers and creating through wars 'giant bazaars for selling the wares of the armaments manufacturers'.[67]

America's power and influence over the international political system was forged through the First World War and the Second

World War. *Pax Americana* is a play-off of *Pax Romana*, the name for a period of concord that extended throughout the Mediterranean world under the rule of the Roman Empire. Imperial rule provided relative stability and peace, as well as some degree of autonomy. Centuries later, during *Pax Britannica*, the British also claimed to be presiding over a period of stability, which lasted until the outbreak of the First World War. Yet like *Pax Romana* and *Pax Britannica*, the *Pax Americana* has been likened to the glorification of imperialism. Analysts such as John Dower have argued that the *Pax Americana* is built on an ever-increasing cycle of militarism that has led to proxy wars, the war on terror, and the build-up of its nuclear arsenal, in each case escalating, rather than deterring conflict and violence. 'The sense that we must always have a dominant military posture means we must always be pushing the frontiers of military technology', Dower observes. 'But that means we are always pushing the edge of greater and greater destructiveness.'[68]

Moyn argues that the *Pax Americana* 'came to the world umbilically connected to air war'.[69] Air supremacy promised not only victory, but also political and economic dominance or control of the United States over other states, their resources and systems, as a mark of hegemony. A concept of control represents a bid for hegemony: a project for the conduct of public affairs and social control that aspires to be a legitimate approximation of the general interest in the eyes of the ruling class and, at the same time, the majority of the population.[70] Those who accept and support America's hegemonic position claim that the United States, as a defender and promoter of freedom and democracy, is in fact a benevolent hegemon. The United States has power and it exerts it, in the Hobbesian that world it inhabits, but only in order to do good. Kristol and Kagan, back in 2000, argued that the United States was a benevolent hegemon, fixing problems like rogue states with weapons of mass destruction, terrorism and human rights abuses. The rest of the world would accept this because the United States and its foreign policy are infused with an unusually high degree of morality.[71] America's benevolent hegemony has been premised on American exceptionalism, 'the idea that America could use its power in instances where others could not, because it was more virtuous than other countries'.[72] Hegemony is more than state dominance. Dismissing America's virtue or strong sense of morality does not reduce its authority to brute force and ruthless exploitation. Hegemony appears as an expression of broadly based consent, involving the acceptance of ideas, supported by material resources and institutions. This may start within the state, then projected out to the world, as with democracy and neoliberalism. Hegemony may be a form of dominance, but it is a consensual one, involving the exporting of ideas regarded as universally good.

Mastanduno argues that hegemony is when a political unit has the power to shape the rules of international politics according to its own interests.[73] Stuart Kaufman, Richard Little and William Wohlforth equate it with hierarchy, the political-military domination of a unit 'over most of the international system'.[74] Ikenberry and Kupchan place the emphasis on material power, most effectively exercised when a hegemon is able to establish a set of norms that others willingly embrace.[75] In a way, a hegemon shapes reality, if hegemony is more than brute force of dominance. Its values, its intersubjective meanings – shared notions about social relations – create a reality, a moral and ideological context, supported by institutions, that shapes desirable and acceptable thoughts and actions. Hegemony filters through economy, culture, gender, ethnicity, class and ideology, so there can be dominance without hegemony, and hegemony is one possible form of dominance.[76]

A New 'Humane and Just' Age?

The United States as a moral hegemon

Writing the War on Terror entails writing identity: the evil terrorists versus the good Americans. The social construction of all war requires an 'othering' process. When he addressed the nation, George W. Bush described the attackers/hijackers as evil, despicable, 'the very worst of human nature', while Americans were 'moms and dads, friends and neighbors', 'strong', a 'great people has been moved to defend a great nation', 'the brightest beacon for freedom and opportunity in the world'.[77] America 'loves peace, America will always work and sacrifice for the expansion of freedom', he declared in 2003, two months after the invasion of Iraq.[78] This was echoed by others. 'Has there ever been a time when the distinction between good and evil was more clear?' asked Krauthammer shortly after the attacks.[79] It was not a distinction that could easily be dismissed or discredited, when we witnessed, live, the murder of thousands of people. How could those men who inflicted such pain, devastation and death, not be evil? And how could those fighting against them, not be good? In this war, even terrorists, once disarmed and detained, would be treated humanely and would be held accountable for their crimes in fair trials. That was two months before the opening of the Guantanamo Bay detention camp, where hundreds of men were to be held indefinitely and without trial, tortured and killed (nine of them died in custody).

Thus started a new type of war, in keeping with a global moral hegemon, that would progressively become not only more ethical, it was claimed, but also more humane. America's 'humanitarian norms' would become conventional wisdom, part of public consciousness, camouflaging an illegal and immoral permanent war and leaving a legacy for making endless war legitimate, rather than ending it. The rise and normalisation of killer drones under Obama was the extension and expansion of the endless war that had started under Bush, relying on no-footprint drones as a new mode of humane killing. American and other Western publics accepted that its humanity made ongoing American war ethically wholesome. In 2016, towards the end of Obama's second term, journalist Jeffrey Goldberg claimed that Obama had become 'the most successful terrorist-hunter in the history of the presidency, one who will hand to his successor a set of tools an accomplished assassin would envy'.[80] The 'arc of the moral universe', Moyn writes, 'ran through the humanization of interminable conflict'.[81] But that was because global terrorism in the twenty-first century was so new and so dangerous that it required new ways of thinking about 'just war and the imperatives of a just peace'.[82] 'Our actions matter', he said in the same speech, 'and can bend history in the direction of justice.'

A hegemon's humanitarian norms and efforts to achieve a just peace? From a realist perspective, states – especially hegemonic ones – do not have such norms or concerns, rather, they are motivated by the desire for power and their own interests. As Mearshmeimer has argued, great powers are rational, not moral actors: they think strategically about how to survive. They operate according to their own self-interest and do not subordinate their interests to those of other states or to the interests of the international community (which realism does not recognise). 'The reason is simple', he argues, 'it pays to be selfish in a self-help world'.[83] States rarely expend blood and treasure to protect foreigners from abuses. State behaviour is driven by calculations about relative power, not by commitment to a world order independent of a state's own interests, or commitment to promoting world peace:

> The particular international order that obtains at any time is mainly a by-product of the self-interested behavior of the system's great powers. The configuration of the system, in other words, is the unintended consequence of great-power security competition, not the result of states acting together to organize peace.[84]

The best way to ensure survival for a state like the United States is to be the most powerful state in the system: the hegemon. As Kant wrote, 'It is the desire of every state, or of its ruler, to arrive at a condition of perpetual peace by conquering the whole world, if that were possible',[85] for if one state achieves global hegemony, the system ceases to be anarchic and becomes hierarchic. If drone warfare helps a great power like the United States achieve global hegemony, control and near-limitless power, while at the same time also *appearing* to be just and humanitarian, then remote warfare is the rational choice.

What also matters to the realist is effectiveness, in terms of protection of its armed forces, as well as in terms of killing the enemy and achieving battle goals. In Iraq, the United States managed to kill as many as 43,881 'opposition fighters'[86] and was eventually successful in getting territory back from ISIS in oil-rich areas, all without putting its military at risk.

There may have been thousands of civilian casualties, but this security perspective regards them as collateral damage. In many cases, civilian deaths came as a result of bombing dual-use targets, that is, targets used both by the military and by civilians during war: roads, bridges, radio and television networks, railway lines and so on. Civilians – those not killed outright – pay a high price for any damage inflicted on such vital infrastructure. The United States did this in Kosovo in 1999 and again in Afghanistan and in Iraq. The military goal of such strikes, reminiscent of realism, is to hit as hard as possible. Children like siblings Lina Wail Mwafaq Mosa Tabra, the fourteen-year-old girl who wrote poetry, and young Mohammed, who always like to go where his sister went, killed as they cowered in a farm compound, bombs dropping on their heads, were collateral damage in America's war on terror on 31 March 2003. Fourteen years later, two more children, sisters Dalal and Amina Ayman Ahmed, were among the dead in strikes that took sixteen lives.

Table 4.12: Iraq Body Count Incident a6262[87]

Incident	a6262
Type	air attacks
Deaths recorded	14–16
Targeted or hit	Al-Aklat flour mill and house hit, casualties include women and children
Place	Bawabat al-Sham, Hay Al-Matahin, west Mosul
Date	14 February 2017

Identifying details			
IBC page	Identifying details (number if more than one)	Age	Sex
a6262-dx3523	Ali Khadr Thanon	47	Male
a6262-ec3650	Aisha Abdel Thanon	43	Female
a6262-dw3465	Hussein Ali Khadr	16	Male
a6262-zx3511	Afrah Ali Khadr	27	Female
a6262-sf3460	Noor Mohammad Hamid	20	*Unrecorded*
a6262-zw3445	Dalal Ayman Ahmed	5	Female
a6262-sb3646	Amina Ayman Ahmed	3	Female

Such are the realist ethics or lack of ethics. What the US strategy shows – which is compatible with realism – is a conventional approach to security that stresses the following:

- enemies/threats must be fought and eliminated;
- threats exist independent of the procedures, discourses and knowledge brought to bear by security agencies;
- security is best served through the use of military power;
- national security is to be prioritised; and

- preparing for war must be the state's principal means of ensuring its own security.

Control and aggression as the means to national security are central to realism. They are achieved with the aid of and recourse to 'Western values', in the context of an imperialist project through which domination is achieved both by military force and by making one's values/ideas/products/lifestyles/systems the dominant ones. Imperialism can take the form of an active, formal and deliberate policy, including or resulting from military action to conquer a state, a culture and its people. It involves extending control over weaker peoples by any and all means: by territorial acquisition, or by the establishment of economic, political and military hegemony over other nations.

Vilmer argues that there are numerous advantages in using drones: clear tactical and ethical advantages that can satisfy even international humanitarian law and counter real-world threats:

> There is a legitimate use for drones in situations of armed conflict, which is no more problematic than that of airplanes and helicopters. That the cockpit is not in the vehicle in the air but somewhere else on the ground does not constitute a relevant difference in most situations.[88]

Drone warfare is different only in that US citizens do not need to shed their blood, which in this case is the primary moral concern of realism. But there is more: what this shows is the power, intelligence and capabilities of humans, as they triumph over their environment. It is the survival of the fittest!

> As animals, human beings have an instinct for self-preservation; and as tool-making animals ... they have always used their ingenuity to protect themselves while killing others. Human capacity for killing at a distance dates back to the Paleolithic era, and always was an engine for the evolution of weaponry (javelins, catapults, bows and arrows, cannons, rifles, revolvers, artillery, machine guns, submarines, airplanes, missiles, drones, and computers).[89]

And anyway, war has always been safer for some. It is a clear reflection of the realist approach and of US policies and actions in the war on terror. As the war in Iraq developed with the appearance of ISIS, the war objectives were still the same, but coalition forces were now lean, nimble and smart. This was the 3-D vicarious warfare of danger-proofing, delegation and darkness, where remote warfare helped achieve the hegemon's regional and global goals. Was it moral? No. But the realist is not particularly concerned with ethics.

The militarist, however, is very concerned with ethics. Like a *Deus ex machina*, the good army enters to deliver justice: to triumph over evil, to punish the bad and to bring moral solutions. In the context of the War on Terror, as the name suggests, the good Americans have been on a mission to not only fight the evil terrorists, but also, at the same time, spread Western ethics: democracy, freedom, fairness, equality, tolerance and respect for human rights. How could a war against terrorism, tyranny, oppression and brutal dictators not be a virtuous war?

At the heart of US military tradition and culture is a policy 'to dominate the world through absolute military superiority and to wage preventive war against any possible competitor',[90] but almost as a mission from God, with all types of warfare being used, including high-tech, to implement a noble world strategy. The standing army created in the wake of the Vietnam war has become sacrosanct. American soldiers are heroes:

> The man who will go where his colors go, without asking, who will fight a phantom foe in jungle and mountain range, without counting, and who will suffer and die in the midst of incredible hardship, without complaint, is still what he always has been, from Imperial Rome to sceptered Britain to democratic America ... As a legionary, he held the gates of civilization for the classical world; as a blue-coated horseman, he swept the Indians from the Plains; he has been called United States Marine.[91]

The United States' geostrategic situation and its military practices can be understood in the context of an American military culture of unprecedented morality, according to which soldiers are regarded as superior to their civilian peers, and apocalyptic mobilisations for war are necessary to counter threats to world peace and justice. American combatants, based on this military culture, have taken a harder and more arduous path from their civilian contemporaries, who lack (their) moral and physical courage. This is an even more moral army of an already moral nation. In this narrative, the United States is a truly global moral hegemon, fighting virtuous wars.

Writing on the military-industrial complex, within which America is waging its wars, Rogers observes that it is rooted in a culture that prioritises military responses to perceived threats, rather than addressing the reasons for revolt, thus making war the 'right' answer, as the complex benefits greatly from conflict.[92] This is where realism and militarism meet: war is both noble and profitable, holding back the 'barbarians at the gate', while pursuing the interests of the armed forces and the military-industrial complex.

But through drones? Where is the warrior's honour? While war is a noble undertaking for militarists, even the 2003 invasion and occupation of Iraq, when American soldiers sacrificed themselves on the 'field of glory', as it is being wholly replaced by remote warfare, those military virtues of courage, heroism and self-sacrifice that make US Marines better than the general population, are disappearing. A moral hegemon has entered a post-heroic age, in which it has become a lot harder to claim moral superiority. The War on Terror started as a total and conventional war that also employed remote warfare, yet within a few years developed into an exclusively remote way of fighting – for the Americans, that is. And while the wars themselves are still considered moral endeavours, fighting them through drones alone, or through delegation, would not be acceptable to the militarist. While lives are no longer lost on the American side, what is lost is the admiration, the pride, the medals, the war memorials, all of which have always been a significant part of a war effort.

As US forces were withdrawing from Afghanistan in the summer of 2021, ending a twenty-year war that had started with the United States invading and occupying the country as part of the War on Terror, a series of suicide attacks on 26 August killed 169 Afghan civilians and thirteen US service members near Kabul airport during an evacuation. One year later, on the anniversary of the attacks, the US Department of State paid tribute to those service members, in a statement titled 'Remembering the Loss of 13 American Heroes':

> One year ago today, 13 American service members lost their lives in a terrorist attack on the Kabul airport in Afghanistan. These American

heroes were engaged in a selfless mission to undertake an unprecedented airlift, making the ultimate sacrifice while helping to successfully move nearly 124,000 people to safety and giving tens of thousands of Afghan families the opportunity to pursue the American dream. We honor their memory and mourn their loss.[93]

The heroism, courage and self-sacrifice of the American soldier, the morality of the American nation and the enviable glory of the American dream are all part of the narrative of the benevolent hegemon. But in Iraq, with the development of 3-D warfare, the only courage shown was by those on the ground: the doctors, the reporters, the parents, grandparents and children, the shopkeepers, the shepherds, the police officers, the actual local combatants, even the terrorists, for they were the ones who ended up making the ultimate sacrifice. The war of targeted killing may have been a smart war, but not an honourable one, as operators 'neutralised' enemy 'soldiers'. Removing risk from the equation, in all-drone wars, puts an end to the very concept of the soldier.

In conventional warfare, it is argued, both sides are morally equivalent. However, militarism would disagree: in this case, the American side would be morally superior. Yet drone warfare takes that moral superiority away, as a result of which the only heroes are non-American, local people, and those that President Bush called 'enemies of freedom'. To make this even worse for the militarist, targeted killing is increasingly carried out by persons who are not part of the military, and the nation's finest are now becoming the fastest clickers. What may be eroded, in the long run, are Western civilisation's mores, as embodied by the world's biggest democracy.

Just War and the Invasion and Occupation of Iraq, 2003–11

The just war theory recognises the distinctive requirements of international relations and war, thus bringing within the boundaries of moral acceptability forms of behaviour that the realist assumes to be beyond the reach of morality. For example, in just war theory strategic bombing is not necessarily without ethical grounding; rather, its morality depends on circumstances, which means that in certain conditions, strategic or precision bombing can be carried out without violating fundamental moral principles. The just war attempt to influence the moral climate of war is not futile. US governments have invested in the development of weaponry that allows wars to be fought in a less destructive, more discriminating and more ethical way, demonstrating a military policy that is susceptible to shifts in perceptions of what is morally acceptable and what is not.

It is useful to examine two phases of US actions in Iraq, while examining moral acceptability based on the just war tradition. Phase one is 2003–11, the period covering the invasion and occupation of the country, until US and UK withdrawal of forces. Phase two is the war against ISIS, 2014–18. Each phase consists of different circumstances and conditions and thus may involve different moral evaluations.

The 2003 invasion of Iraq was an act of aggression, the unprovoked attack by the global hegemon and its powerful ally of a small, weak state on the other side of the world, resulting in thousands of civilian deaths in the space of a few weeks. There followed eight years of occupation, during which Coalition commanders established permissive 'rules of engagement': swift and unhesitating use of force to maximise impact and to minimise their own casualties. Rules of engagement define when, where and how military personnel can use force. Commanders used 'kill counts' and other devices that encourage competition among soldiers to rack up 'enemy kills'. The result was a rapid escalation of force, leading to large numbers of civilian casualties. An environment of extreme violence produced killings and atrocities, including war crimes, committed by Coalition forces against Iraqi civilians.

Troops frequently opened fire at checkpoints (often set up suddenly and unexpectedly, at night or on roads with poor visibility) causing excessive and unnecessary deaths of approaching civilians, on foot or in vehicles. One such victim was Walid Fayay Mazban, who was driving with his family in Basra on 24 August 2003:

> He was the sole breadwinner for his wife, two children and two parents. On the evening of 24 August UK soldiers were staffing a temporary checkpoint at the Suq al-Hattin crossroads on the edge of Sikek. According to an eyewitness, three soldiers stood across the northern side of the road, one was on the southwestern side and another was on the northeastern side. The area was very dark because there was no electricity ... He was driving normally and suddenly there was gunfire. It was dark. There was no light.[94]

Walid Fayay Mazban was fired on six times from behind, as he drove, and died of his injuries.

Coalition forces routinely searched houses, using disproportionate violence to break in, such as shooting door locks, placing a bomb or hand grenade outside the door and battering down front walls with military vehicles. Once inside, troops would 'prep a room' by spraying it with gunfire.[95] In Haditha, two house-searches resulted in the deaths of fifteen people.[96] During patrols, coalition forces would regularly fire on Iraqis. Based on estimates by the Iraqi Police in Baghdad, US forces killed thirty-three unarmed civilians in the capital alone, between 1 May and 12 July 2005.[97]

On 19 November 2005, a squad of US Marines went on a rampage after a roadside bomb killed one of their group in Haditha. The squad's leader killed five unarmed young men who happened onto the scene in a taxi. The marines then raided nearby houses, killing twenty-four civilians, including children and an elderly man in a wheelchair.[98] The incident was documented by Iraq Body Count and all twenty-four victims were named, from seventy-six-year-old Abdul Hameed Hasan Ali, to two-year-old Aisha Younis Salim.[99] Like the torture in Abu Ghraib, US officials first described the Haditha massacre as an isolated case of misconduct, but further revelations showed that it was part of a pattern of unrestrained violence common among coalition troops.

Air attacks, increasingly using killer drones, rose five-fold from January to November 2005.[100] In one typical week in March 2007, the Pentagon reported 327 missions, or about forty-eight every day. Despite the use of sophisticated, precision-guided munitions, civilians were routinely killed and injured.

The most damning official document, however, came in the form of the 2020 International Criminal Court report on British war crimes in Iraq. According to the report, British soldiers committed more than 1,000 acts of torture, sexual violence and unlawful killings against male detainees in Basra, between 2003 and 2008. 'Such mistreatment was systematic and had a systemic cause, which further suggests that there are hundreds more such victims.'[101]

This was not a just war.

In addition to the invasion killings, arbitrary detentions, extrajudicial killings and torture, reprisals and other heavy-handed measures are not just evidence of an unjust war, but closely resemble acts committed and measures taken by the Taliban in

Afghanistan in 2022, as detailed in a Crisis Group Asia report: arbitrary detentions, extrajudicial killings and torture, evictions and reprisals, profiling and collective punishments, house searches and raids.[102] The invasions of Afghanistan and (even more so) Iraq were unprovoked, aggressive and catastrophic, with extreme brutality shown and long-term consequences, engendering insurgency and terrorism that killed tens of thousands more.

The now (in)famous and largely acceptable – yet immoral – pre-emptive strikes were, in the words of Francisco de Vitoria, like killing a man for a sin he has yet to commit. A just war cannot be grounded on imperial expansionism, nor can it be undertaken unless all non-violent alternatives are exhausted, because the consequences of war are cruel and horrible.[103] Cicero, in a discussion on the laws of warfare, also wrote of unjust wars, waged for supremacy.[104] Just war theory cannot justify wars waged to serve the interests of powerful states in the context of colonialism or neo-colonialism. Non-Western people hold equal legal and moral status with Western (European and American) people, regardless of differences in culture or in power. The right intention is a moral condition for waging war, and 'one polity's interest in the assets of another polity, or its wish to dominate another polity, cannot contribute to the moral justification of a war'.[105]

Neither near, nor remote warfare can be morally justified in this type of war.

Just War (2) 2014–18

Was remote warfare just in the war against ISIS? It can certainly be argued that this was a just war. The 2014 ISIS invasion and occupation of Iraq was not unlike the 2003 war the Coalition had waged, in terms of its brutality and boundless force. The first mass killings of civilians were not unlike 'Shock and Awe': between 12 and 15 June, the group abducted 1,566 unarmed air force cadets from Camp Speicher in Tikrit, tortured and executed them. The Camp Speicher massacre was filmed, photographed and publicised widely by ISIS fighters, in an attempt to further terrorise and intimidate. Less publicised, but no less shocking, were subsequent executions of Iraqi civilians. A search of the Iraq Body Count database during this period reveals the extent of these massacres, as well as the presence of mass graves. Below is the result of a search using the 31 August 2014 date. It reveals this horror:

Table 4.13: Massacres and mass graves by ISIS[106]

a0350c	80–100 bodies of Yazidis found in mass grave in Sinjar, west of Mosul	1 Aug 2014 to 31 Aug 2014
a0350d	16–102 bodies of Yazidis found in mass grave in Zummar, west of Mosul	1 Aug 2014 to 31 Aug 2014
a0350f	80 bodies of Yazidis found in mass grave in Badoosh, west of Mosul	1 Aug 2014 to 31 Aug 2014
a0350e	23–103 bodies of Yazidis found in mass grave in Zummar, west of Mosul	1 Aug 2014 to 31 Aug 2014
a0350j	250–300 Yazidis killed in Hardan village, northeast of Sinjar	1 Aug 2014 to 31 Aug 2014
a0364	Bodies of three men found in Tuz	1 Aug 2014 to 1 Jan 2015
a0386	Journalist executed in Ghizlani Camp, south-west Mosul	1 Aug 2014 to 20 Jan 2015
s0490	Mother and daughter by shelling in Muslakha, 40 kilometres north of Tikrit	31 Aug 2014 to 1 Sep 2014

The group is by now best known for the genocide of the Yazidi population of Iraq, against which they committed a variety of violent crimes. Thousands of adults and children were killed, either executed or shot as they fled. Many died on Mount Sinjar as they tried to escape. Thousands more were enslaved, with women and young girls subjected to serial rape, sometimes leading to their deaths. Numerous other international crimes were also committed against the Yazidi community, 'including extermination, enslavement, sexual violence, forcible transfer, persecution on religious and gender grounds, and conscription of children into an armed group'.[107] The torture and execution of thousands, as well as outrages on their personal dignity, are acts that constitute was crimes.

A war against ISIS then can be justified as a just and moral war, which allows us to move from *jus ad bellum* to *jus in bello*. If the war is just and it takes place within the context of ISIS brutality, massacres and genocide, then it seems that, based on just war principles, the use of drones would be morally acceptable, even if the drones were not always precise and killed a great number of civilians too (questioning any claims to proportionality) – even more than they had killed during the 2003 invasion. Just war would recognise right intention in trying to defeat a terrorist group that is committing such crimes in areas they are occupying, and to bring peace and respect for human rights in war-torn Iraq. Burkhardt writes that a state with right intention 'establishes conditions for a just and lasting peace by respecting human rights, taking due care to insulate civilians from the harms of war, allowing for political self-determination, and by educating its military and political culture'.[108] The United States does not quite fit this description, because in Iraq the devastating casualties from drone strikes did not exactly demonstrate great respect for human rights, but it could still be argued that remote warfare, by using limited force, was less destructive than a conventional, large-scale war would have been. Moreover, based on the nature of this particular terrorist group, an armed response could be seen not only as urgent, but also as necessary.

Conclusion

In March 1944, while the Second World War bombings over German cities were ongoing and the worst bombing in history was yet to come, *New York Times* editorialists wrote, 'Let us not deceive ourselves into thinking that war can be made more humane. It cannot. It can only be abolished.'[109] In other words, war cannot be cleansed of its crimes, because war is the biggest crime.

Through the development and use of new technologies, there have been attempts to make war more humane or ethical. Have those attempts failed?

The three perspectives give different answers, with realism being the only clear supporter of remote warfare through the use of killer drones. The case study of the War on Terror and America's use of UAVs highlighted the realist emphasis on the protection of armed forces, the primacy of great powers and the insufficiency of civilian deaths as a condition for right action. If the killing of civilians is necessary for victory, if the drones are less precise than is desired, if thousands of Iraqis/Afghans/Syrians die along the way, the realist would still support drone warfare as more moral and certainly more effective. For realists, security is examined and understood at state level, and it is at state level that the moral evaluation is made.

From a militarist perspective, drone warfare alone would not make for a moral or heroic war. As for just war, under certain conditions – for example, in America's war against ISIS – a case could be made for *jus ad bellum* and *jus in bello* using drones. Yet questions remain, regarding both. *Jus in bello* is made complicated by the fact that, although necessity may be clear, proportionality presents a challenge, due to the extremely high number of civilian casualties, that may just be too high to consider this war ethical or just. The contradictions in US foreign policy are at the heart of this problem: while policy initiatives purport to promote peace, stability and human rights, the consequences are devastating, immoral and arising from hugely violent actions. The UA position seems to be that people in other countries may be sacrificed during missions to protect US interests. Aided by what could be called an ideology and practice of technological Darwinism, few US advocates appear to register that the targeting of suspects using drones would be denounced as unconscionable crimes of aggression, if they were carried out by a foreign government on US soil. In the War on Terror, a hybrid war-law model came to govern America's use of force, especially against non-state actors, based on which it claims to possess the right to wage war against terrorist groups, while adapting the logic of law enforcement that denies suspects the right to fight back, or to have the legal status of prisoners of war. The war-law model both justifies the use of lethal force and eliminates the rights of adversaries and civilians. Just war theory may seem overly permissive in this context. Just war principles are problematic when it comes to limited force, in a case such as this, and may even provide moral justification for acts where greater restraint should have been exercised.

Is it even possible to discuss the ethics of warfare – even the remote type – outside the ethics of politics? The use of force in war does not take place outside politics. Clausewitz argued that war does not suspend or change political intercourse: military events progress along political lines, and during the war the logic of politics continues in its most profound manifestation. This means that military events must be understood in their political contexts and can be judged only according to the political effects that they are trying to achieve. Military operations are embedded within what Waldman calls a political web of war; therefore, we must always seek to understand how interests, relationships and political dynamics are being shaped by military moves – and how they (rather than humanitarian concerns) influence the decision to use military force. With that in mind, even America's decision to start a war against ISIS in 2014 is questionable, in terms of ethics; the *jus ad bellum* may actually not be quite so just, but interest-led, understood within the wider neo-imperialist war on terror, a war waged to control, to exploit, to secure resources and to eliminate all threats to America's gains.

Finally, the issue that remains is that even those who are targeted have their rights and must be presumed innocent. That they are a threat to someone does not make their deaths morally acceptable.

5 • Remote Killing and the War in Ukraine

> What will bring the end of the war? We used
> to say 'peace'. Now we say 'victory'.
> PRESIDENT ZELENSKYY SPEECH ON 24 AUGUST 2022[1]

> Stranger, make this message clear to those at home who hold us dear.
> We honoured what our country said: to hold
> our ground – and here lie dead.
> FROM THE EPITAPH FOR THE SPARTANS
> WHO FELL AT THERMOPYLAE IN 480 BCE.[2]

The War on Terror has shown how remote warfare both stretches and shortens the battlefield, as United States launched risk-free drone strikes in Afghanistan, Iraq, Pakistan or Yemen, in mostly secret and targeted attacks. Those 'precision' targeted assassinations sometimes hit the targets – those individuals they hunt – and sometimes they don't, with devastating consequences. The ethics concerning the ability of a global hegemonic power to kill – and to do so with complete impunity – citizens of other states who pose a threat to its interests, creating perpetual grey zone conflict, were examined and evaluated through the application of different ethical approaches to warfare and security. The ways in which drones are being used in the war in Ukraine, however, appear to be different from the ways the United States has deployed them. In Ukraine, both sides use drones for a range of missions, including 'battlefield surveillance, artillery spotting and attacking armored vehicles and missile launchers'.[3] The Ukraine-Russia conflict is carried out on a hypermodern battleground where, more than ever before, drones foreshadow a world in which armed conflicts could end up being conducted exclusively by remote control and even fully by artificial intelligence.

In many ways, the war in Ukraine resembles past conflicts that involve the use of fighter planes, tanks, helicopters, anti-aircraft missiles and warships. Yet in addition to more traditional warfare, the Russian invasion of Ukraine has led to a high intensity war 'where both sides have extensively deployed military and commercial drones as an extension of air power or as ammunition'.[4] In this war, armed drones have not only been used extensively, but are also more politically acceptable, especially when used by the Ukrainian side, which is seen as fighting a defensive war. This chapter will then begin by looking at the morality of self-defence killing and national defence, a discussion that will prove particularly useful when it comes to a moral appraisal through the lens of just war, which holds that a defensive war is a just war. The chapter will first assume that the war the Ukrainians are fighting is a defensive war and, within that context of defence, discuss weaponry, impact and ethics. It will then query this assumption and argue that this is not a defensive war, following which the ethics of *jus ad bellum* and *jus in bello* will be revisited.

Self-defence and National Defence

The ethics of self-defence have implications for the moral justification of national defence, whereby self-defensive behaviour, whether by individuals or by nations, is morally superior to aggressive behaviour. Self-defence has an 'aura of respectability that makes it eminently suitable as a rallying point for agreement on the ethical legitimacy of warfare'.[5] Killing a person in response to a threat to one's life, as in defensive warfare, is regarded as a morally justified self-defensive act: killing in self-defence when that is the only option available for saving one's own life is morally justified lethal self-defence. On the other hand, killing when it is done by aggressors is not morally justified; the aggressor's behaviour is morally wrong, while the victim's (aggressive) response is morally right, because of the victim's moral right to protect himself/herself from deadly aggression. There is then a presumption of the aggressor's culpability and the victim's innocence/lack of culpability. According to Hobbes, a human being's natural propulsion towards self-preservation requires them to kill, when they are at risk of being killed. Which means that it is natural and (at the very least) morally permissible for humans to kill in self-defence.

It is presumed in arguments justifying self-defensive killing that the victim – or perpetrator of the self-defensive killing – is not responsible for bringing about the threat to their own life. The relationship between an aggressor and a victim is regarded as asymmetrical, in that the aggressor is fully responsible, whereas the victim is not at all responsible for causing a situation in which the victim has no choice but to threaten and extinguish the life of the aggressor. A victim incurs no responsibility for having killed an aggressor who threatened his or her life, but, on the contrary, makes the aggressor take responsibility for his or her actions, by letting them bear the consequences of their actions. As long as victims are placed by their aggressors in situations where they must choose between killing their aggressors and being killed by them, they have no moral obligation to refrain from killing their aggressors.

The principle of morally justified self-defensive killing in war implies that aggressive enemy forces in uniform can be killed by other combatants/soldiers fighting a defensive war, unless they are surrendering or are *hors de combat*, because the lives of the soldiers defending their country and the lives of their co-nationals are threatened by the enemy forces. In this case, which is the case made for Ukraine, the imminent threat to life of those on the battlefield can only be averted by killing the enemy. Morally, defensive military action is very different from military aggression, with the main goals of defensive military action being to ward off aggression, to save lives and to limit the bloodshed, all through the use of force.

However, there are obvious problems – the first of which is that, in the case of war, it is rarely clear what is meant by 'victim'. Is the victim the soldier, is it the military leadership, or is it the political leader of the state that is being attacked? Depending on whom we take to be the 'victim', we would give a different response to the question of whether they could have acted otherwise, perhaps in non-lethal ways. The second problem is that if the aggressor's behaviour is morally wrong, it does not necessarily follow that the victim's behaviour is morally right. The third concerns the victim's responsibility or contribution to bringing about a conflict where kill-or-be-killed becomes the norm. Finally, is killing human beings a morally right response, if what is threatened is the taking away of one's freedom, power, status or rights? Following a declaration of war by a state, even one resulting in an invasion and occupation of another state, there may be options for non-lethal resistance. Those whose country is under occupation are confronted with a fate that is not as bad as being killed, and so killing the invader is not necessarily – or at all – a morally justified response. It is doubtful whether using deadly force to protect rights and freedoms is morally justified.

As for the question of who is responsible for bringing about the threat of war – the basis on which a determination can be made on what warrants the killing of others – there is no single or simple answer, but a multitude. In his book *On the Causes of War*, Suganami suggests four types of acts-as-threats: acts with belligerent intent; insensitive acts; thoughtless acts; and reckless acts.

When one state confronts another explicitly with a choice between immediate surrender and war, then such an act is said to be undertaken with 'belligerent intent'. Such acts, Suganami points out, are not necessarily aggressive, for it is possible for them to be undertaken in response to varying degrees of hostile provocation. Suppose that state A perpetrates what amounts to an offensive act against its adversary state B – an act whose offensiveness would be apparent to a reasonably attentive mind. If A performs the act unaware that B will, with good reason, consider it offensive, then A can be said to act 'insensitively'. However, if A performs the offending act, having noticed its offensiveness, but failing to give any serious thought to the resulting risk of war, then A can be said to act 'thoughtlessly'. Finally, A may come to calculate as follows:

> The probability of war resulting from the act, and/or the cost of such a war, might be considerable; but still the probability and/or the cost would be tolerably low when judged in the overall context of what A could probably gain by resorting to the act in contrast to not doing so at the time.[6]

An act carried out after such deliberation, opting for the risk of war as the more beneficial option, is a 'reckless' act.

This discussion on defence, responsibility, gain, acts-as-threats and causes of war will become pertinent in the moral evaluation of the war and the way that it is fought in Ukraine.

Invasion of Ukraine by Russia and Ukraine's War of National Defence

In *Invasion: Russia's Bloody War and Ukraine's Fight for Survival*, Harding adopts a self-defence approach when narrating and analysing the war. The clear aggressor and instigator of the war is an imperialist Russia. He writes:

> Putin had issued a series of demands so imperious and swaggering you could only marvel at their audacity. He sought nothing less than the annulment of the security infrastructure that has governed Europe for the three decades since the Soviet Union's 1991 collapse. Further, he wanted the Biden administration to guarantee Ukraine would never join NATO, the United States-led military alliance set up in 1949 to contain the Soviet Union.[7]

Before invading Ukraine on 24 February 2022, Putin also demanded that NATO take its forces and equipment out of Romania, Bulgaria, Poland, Latvia, Estonia and Lithuania, former Eastern Bloc countries that had joined NATO after 1997. Putin's goal, Harding argues, was to re-create the Soviet Union's sphere of influence that had existed across the European continent behind the 'iron curtain'. This area 'encompassed Belarus and Ukraine – "historic" Russian lands, as Putin saw them – unjustly separated from Moscow by Bolshevik blunder and Western meddling'.[8]

Putin used tactics familiar from Russia's dark past, both distant and recent: bombs, destruction and the killing of civilians. This time Russia's direct enemy was Ukraine, its indirect enemy its Atlanticist leadership. In Russia's attempts to create a new world order and become a global hegemon, this invasion would become the largest conflict in Europe since 1945: 'an attempt by one nation to devour another'.[9] One expert even described it as Russia's strategy to wipe out 'Ukraine-ness',[10] while simultaneously fighting a proxy war against the West. And it was a West-backed Ukraine that responded to the Russian aggression. Within days of the invasion, Sweden and Finland abandoned neutrality, Germany announced a radical shift in its security policy, and the United States and its allies found a new role and moral purpose in their solidarity – both moral and material – with Ukraine.

Yet it was also a (defensive) war agenda that the words of Ukraine's president betrayed from the start. Zelenskyy's presidential term began on 20 May 2019, almost three years before the invasion. On the same day, referring to the Russian annexation of Crimea and the Russian-instigated conflict in Donbas, in his inaugural address to the Ukrainian Parliament, he declared: 'I am ready to pay any price to stop the deaths of our heroes' and secure 'the "return" of the lost territories ... Crimea and Donbas are Ukrainian land.'[11] He soon embarked on a quest to find willing and powerful allies to help him defend Ukraine.

On 25 September 2019, Zelenskyy addressed the UN General Assembly in New York: 'What is happening in my country is no longer "someone else's war". None of you can feel safe when there is a war in Ukraine. Not when there is a war in Europe ... Military methods, technologies and weapons mean that our planet is not as large as it once was', he pointed out, adding, 'A strong leader is the one who protects the lives of everyone.'[12] In his address to the United States Holocaust Memorial Museum, Washington DC, on 1 September 2021, he urged, 'Do not be indifferent to Ukraine.'[13] Zelenskyy's speech to the Munich Security Conference, 19 February 2022, made it clear that 'We appreciate any help, whether it is hundreds of modern weapons or 5,000 helmets.'[14] The day after Russia invaded, Zelenskyy addressed the people of Europe from Kyiv: 'I know that Europe can see this. But what we do not see – at least not fully – is what you are going to do about it ... to protect Ukraine.'[15]

Appearing remotely in the German Bundestag, three weeks after the invasion, he urged Germany to 'Support us. Support peace'. His contention was that Ukraine had always striven for peace, and had always been committed to dialogue, to negotiations and to a diplomatic solution. Yet on 24 February 2022, at 4:30 a.m., Ukraine had received Russia's answer, made clear through its actions: Russia wanted 'to destroy Ukraine; to wipe us off the face of the Earth, both as a state and as a people'. Instead, though, months into the invasion, the Russians saw 'the kind and peaceful people of Ukraine turn into lions, ready to tear apart any enemy'.[16] 'They have watched a friendly and hospitable people become warriors.'[17]

The impact on Ukrainians – the warriors and the non-combatants – has been well documented. From 24 February 2022, which marked the start of the large-scale armed attack by

Russia, to 18 June 2023, the UN Human Rights Office recorded 9,083 civilians killed.[18] Reminiscent of the Iraqi dead, the names of those lost were entered onto an online memorial, trying to contain the immense loss of a mother, a grandmother, a father, a son, inside a few words:

> Oleksii Bondar was killed on March 19, 2022, in Mykolaiv. The Russian army carried out an airstrike on his family's house. The 7-year-old boy died with his mother and grandmother. Oleksii had just started school. He liked to play football.[19]
>
> 22-year-old Mykola Klontsak was killed on March 11, 2023, in the battle for Bakhmut. 'He had been working since he was 14. He grew up without a father, his father died when Mykola was 11 years old. I always knew I could count on him. Now he is our Angel,' – says his mother.[20]

Many of those now fighting – and dying – in defence of their country were civilians, when the war had started:

> 22-year-old Vasyl Kosovskyi was killed in action on March 16, 2023, in Donetsk. Vasyl lived in Sumy Region. He worked as a milk truck driver and a car mechanic.[21]

Much like the wars in the Middle East, this war has already taken tens of thousands of lives, with more than 100,000 estimated killings of civilians and combatants, Ukrainians and Russians, in sixteen months.

Remote Weapons

On 31 December 2022, Ukrainian quadcopters buzzed over the town of Bakhmut, dropping bomblets on Russian soldiers across the battlefield. Equipped with thermal imaging cameras that helped them identify targets in complete darkness, the small drones managed to fly undetected, while in the Kyiv Oblast region, Russian loitering munitions struck Ukrainian energy facilities and apartment complexes, killing up to four people and psychologically terrorising the population.

Drones have taken centre stage in the Russia-Ukraine war, with Ukraine building an 'army of drones', incorporating small as well as civilian drones, modified for military use. Early in the war, Ukraine used the Turkish-manufactured TB2 Bayraktar to sink the Moskva (the flagship of the Russian Navy's Black Sea Fleet) on 14 April 2022; just weeks earlier, 1,000 Kamikaze Switchblade drones – referred to as 'loitering munitions' – had arrived from the United States, as part of its assistance packages worth $40 billion.[22] The United States has also supplied Ukraine with ALTIUS-600 drones, designed to gather intelligence and to kill. The ALTIUS-600 small unmanned-aircraft system (UAS) can be deployed from the ground, fixed-wing aircraft, helicopters, and maritime platforms, and land on any flat terrain. 'The UAS' flight can be manually controlled by its operator with a handheld remote control system or pre-programmed by a ground control station to operate autonomously.'[23] A high explosive warhead can be fitted to strike targets of opportunity. In addition, armed forces have been receiving drones in large numbers through 'dronations' from civilians, including via crowdfunding campaigns. These drones have been repurposed for spying and dropping explosives on targets. More recently the United Kingdom provided Storm Shadow long-range missiles that were used on 8 June 2023 to destroy a Russian military base in Luhansk Oblast.

The war has provided a testing ground for foreign drone powers, accelerating the proliferation of drones. These are not the kind of drones used by the United States in the War on Terror in Afghanistan, Iraq, Pakistan and Syria, where the airspace is largely uncontested, but much smaller and better suited to an active battlespace. While large drones can be destructive under conditions of air superiority, small drones are becoming crucial for battlespace awareness of infantry and manoeuvring units. Moreover, low-cost, 'kamikaze' attack drones offer yet another way to deliver explosives. In Ukraine's contested airspace, small drones have changed the operational tempo of artillery, shortening time-critical targeting and firing cycles to three to five minutes.[24] Scout drones provide situational awareness and real-time view of the battlefield, which enables soldiers to spot enemy positions without risking the lives of special forces. Small drones have increased 'precision and pace of artillery fires and keep soldiers safe'.[25]

Kamikaze Switchblade drones ('loitering munitions' or 'suicide drones') are small, non-recoverable attack drones that detonate on impact. They can fly up to 160 kilometres an hour and can zero in on targets up to 10 kilometres away. A Kamikaze drone behaves like disposable ammunition that can loiter in the target zone prior to impact. Kyiv also received American-made Phoenix Ghost and Polish-developed Warmate drones. On 29 October 2022, Ukraine launched an attack on the Sevastopol Naval Base housing Russia's Black Sea Fleet with drones and kamikaze 'drone boats' – uncrewed surface vessels – filled with explosives.[26] The R18, developed by the Ukrainian Aerorozvidka, is an octocopter (a vertical take-off and landing drone with eight propellers) specifically designed to drop bomblets. It has a range of 4 kilometres, it can fly for forty minutes and can carry 5 kilograms of free-falling bombs, which it can drop from a height of 300 metres, while hovering over the target. It can fly without lights in the dark.

Both Russia and Ukraine have deployed loitering munitions. Russia has made great use of Lancet-3M or Product 52, that weighs about 15 kilograms and cruises at about 70 miles per hour. The operator identifies targets through the drone's camera, then the drone dives at 190 miles per hour to impact with an armour-piercing warhead. Russia has also been employing Iranian-supplied Shahed-136 kamikaze drones. Renamed Geranium-2, these long-range loitering munitions carry 50 kilograms of explosives over 2,000 kilometres, and used together with cruise missiles, destroyed a third of Ukraine's electrical grid. Millions of Ukrainians lost power in the midst of winter, and critical infrastructure suffered serious damage. When used against civilians, as in Russia's October 2022 strikes on Kyiv, low-altitude drones can not only kill and destroy, but can also terrorise entire cities.

Missiles and drones now dominate the air war over Ukraine, littering the day and night sky. Below, people scramble for cover. The war has largely been fought inside Ukraine, though Moscow has reported some attacks on its territory, which means that so far only Russian drones have killed civilians.

On 3 May 2023, Radio Free Europe/Radio Liberty's (RFE/RL) Ukrainian service reported Russian drone attacks that killed twenty-one Ukrainian civilians in various areas, including the Kherson region.

Ukrainian officials said 21 civilians were killed when various civilian targets in the southern region were hit. (They) all died in Russian strikes that hit a supermarket, a train station, residential buildings, a hardware store, and a gas station in the Kherson region. 'As of now, 21 people have died! 48 wounded! All civilians! In one incomplete day! In one area!' Zelenskiy wrote on twitter on May 3, vowing to defeat Russia and hold all perpetrators to account.[27]

The drone strikes lasted an entire night and day. Among the dead civilians were three employees of an energy company, killed as they worked. The Ukrainian Air Force Command destroyed twenty-one (out of twenty-six) Shahed-136/131 drones that had been launched in this attack, from Russia's Bryansk region and from the eastern coast of the Sea of Azov.

Ukraine's Drone Wars: A Realist Justification and a Militarist Critique

There are two starting points or initial assumptions, in this evaluation:

1. Ukraine is fighting a defensive war.
2. Ukraine kills only Russian soldiers.

The realist's evaluation of Ukraine's remote warfare would take into account a number of factors, starting with the claim that this is Ukraine's fight for survival. In national defence killing, state forces can kill aggressive enemy forces by any means. When attacked, war is a necessity and *jus in bello* is limitless, for realists. For Ukraine's security, the protection and pursuit of its national interest are paramount, and defensive war in this case best serves its political ends and ensures its survival.

In addition, the use of drones such as loitering munitions protect Ukrainian forces mentally, psychologically and materially. In a situation of 'kill-or-be-killed', 'my-state-or-your-state', 'my-security-or-yours', a state must have more and better weapons, more effective, precise, destructive, lethal drones – and not only during active war, but also in peacetime, because war is both necessary and inevitable.

Without war, for realism, there would be no states, because states are born of war. Yet they can also cease to exist through war, so the most important task for a state like Ukraine is to protect itself: its nation, institutions, property, armed forces – its very Ukraine-ness. The more and better Ukraine is armed, the more effectively it uses its war technology, the greater the advantage over its adversary, the better it can ensure its survival and security, and the more it can protect its armed forces. Given that tens of thousands of Ukrainian soldiers such as Oleksandr Pylypets, combat medic Olena Bezkrovn, Danylo Melnyk and Yuriy Taranukha, have been killed so far, increased use of drones could only be good, as drones would prevent such deaths, keep the armed forces intact and strong, and help Ukraine with fewer losses.

However, realism does not distinguish between good and bad states, but between powerful and weak ones. Moreover, realism favours and legitimises the actions of powerful states. All states ought to do what's in their best interests, and the most powerful ones have the greatest capability to do so, which makes it legitimate for powerful states to exert influence and force over smaller and weaker states. Realism would justify both wars – Ukraine's and Russia's, the defensive and the aggressive – but especially the one being fought by the greater power. As a result, in the case of the Shahed drones that killed twenty-one Ukrainian civilians on 3 May 2023, including the three employees of the energy company, those killings would not be judged as immoral, unjust or illegal, but regarded as collateral damage.

As long as drones protect the lives of state forces and increase the state's chances of survival and security, their proliferation is justified and recommended, especially for powerful states like Russia, with imperial designs and a long history of aggression.

While for the realist war is chosen on pragmatic grounds, militarists are proponents of war. In militarism, a military response is a moral response, and it is the preferred response. There is an inclination to rely on military means for the handling of conflicts. This is the response of heroes. Militarism (as discussed in Chapter 1) is the propensity to use military power, or the threat of it, for political settlements. That Zelenskyy chose to fight Russia, rather than surrender, or compromise, is important for militarism. Zelenskyy's choice of military or military-like attire, since the start of the war, fits with this narrative of the brave warrior nation, as do his statements regarding the courage and fighting spirit of the Ukrainian people. In the years leading up to the 2022 invasion, militarism became embedded within society, and the response to the threat of invasion highlighted the carrying of military norms into the civilian sphere.

Since militarism is an ideology that glorifies war, military institutions and the prevalence of martial values, Ukraine's response to Russia was the moral response; the defensive war that resulted was a moral war; its casualties a reminder of the Ukrainian soul. And while the military build-ups, especially with quantitative increases in weapons production and imports, including drones, are in keeping with militarism principles, the war remains virtuous as long as drones remain ancillary weapons. A fully 'dronised' war is not desirable. For militarists, war has ethics and it has heroes, like soldiers Pylypets, Bezkrovna, Melnyk and Taranukha. Like Bondarenko and Miziak, young men who also died for Ukraine:

> 22-year-old Vadym Bondarenko died on the night of May 26, 2023, while on duty. Vadim was born in Cherkasy. He loved to play football since childhood and even played for a local team. At a young age, Vadym was raising 3 daughters with his wife Anna.[28]

> Vladyslav Miziak was killed in action on March 30, 2023. He was 27 years old. Vladyslav was born in Kharkiv Region. Since childhood, he dreamed of becoming a soldier. 'He was an extraordinary person' – wrote his girlfriend Tetiana.[29]

While the whole Ukrainian nation is a nation of heroes, it is the soldiers who are the most extraordinary people. And while the Russians can and do make similar claims about the Russian nation and its soldiers, in Ukraine's case of defensive war these brave warriors, as in ancient Thermopylae, 'fell' defending their country. Ukraine's national anthem contains and repeats these lines:

> Soul and body shall we lay down
> For our freedom.

There is nothing noble, moral or courageous about a kamikaze drone. The mere use of the word 'suicide' to describe the function of such a drone, as though the machine had somehow died, would be considered at worst offensive, at best embarrassing, for the militarist. What is the Switchblade drone compared to Vadym Bondarenko, who risked and lost his life fighting so that his young daughters would live in freedom? What are 'drone boats' next to the extraordinary Vladyslav Miziak? And what is it that the machines have sacrificed?

A Just War of Self-defence

In terms of *jus ad bellum* and *jus in bello*, the war being fought by Ukraine appears to be just for at least four reasons.

First, the war is defensive and in just war theory a defensive war is a just war. The second reason concerns the principle of distinction. While Russian weapons – including drones – kill Ukrainian forces and civilians (such as those killed in Kherson on 3 May 2023), those used by Ukraine have – so far – killed only Russian forces. Armed forces are legitimate targets. The third and fourth reasons address the principles of proportionality and intent. In just war the desired end should be proportional to the means used, a principle that overlaps into the moral guidelines of *jus in bello*. With regard to just cause, a policy of war requires a goal that minimises the war's destruction:

> If nation A invades a land belonging to the people of nation B, then B has just cause to take the land back. According to the principle of proportionality, B's counter-attack must not invoke a disproportionate response: it should aim to retrieve its land and not exact further retribution or invade the aggressor's lands, or in graphic terms it should not retaliate with overwhelming force or nuclear weaponry.[30]

Ukraine (in this example, nation B) has just cause to defend its land and, as things stand, has shown no desire to do anything other than take ownership of land that it sees as belonging to Ukrainians. There is no indication that Ukraine intends to retaliate with overwhelming or nuclear force. The increased use of drones, in fact, indicates the opposite: that Ukraine's goal is to keep casualties low and to expel the aggressor, not to escalate, seek revenge or target Russian civilians.

However, there are problems with this account. First, the conduct of any war, even on a just side, presents moral dilemmas, contains instances of injustice and leads to violations of human rights, cruelty, violent deaths and immeasurable human losses, which means that there is no clean just warfare. Even a just war is only barely just. In all war, children are 'torn from the embrace of their parents, matrons subjected to whatever should be the pleasure of the conquerors, temples and houses plundered, slaughter and burning rife … all things filled with arms, corpses, blood, and wailing'.[31] War creates moral conflicts for all who participate in it and has horrific consequences. 'Since the horrors of war are so morally repugnant, and there is practically no way to keep one's hands clean while fighting, it seems that avoiding them must take precedence over many interests, and even many duties.'[32] The only way to avoid carnage, death and loss is to avoid war. As Miščević puts it, there is a 'very strong and urgent duty of everybody to work on prevention of war'.[33] The duty to avoid war, even defensive war, it can be argued, is a fundamental political and moral duty. The forces that war unleashes cannot always been predicted or controlled – there is too much 'fog', to use Clausewitz's term. Once unleashed, a war may have repercussions that last for generations.

According to just war theory, for the reasons mentioned, war should always be a last resort: all other possible solutions must have been attempted to avoid war. Crucially, it is the responsibility of political leaders – the proper authority – to explore and pursue solutions other than war, as it is the relationship between a government and its people that dictates this duty, this imperative to protect those they lead. What can be justly defended through loss of life? Freedom? Territory? Property? National pride? If compromise and even loss of territory could have prevented a war that has led to the deaths of more than 1,000 Ukrainian children and babies, wasn't peace the better option? Is war still morally justified? In other words, if Zelenskyy chose war and ordered Ukraine's armed men to kill and to risk their own lives; if he chose war and placed at risk of death the lives of millions of Ukrainian civilians; if he chose war when what was threatened was freedom, power, status or rights (not human life, in a clear-cut kill-or-be killed scenario), when he could have chosen peace, would just war theory still defend his decision to go to war? Was not his moral responsibility to protect his people by avoiding war, rather than to build up Ukraine's military, by spending more than two years (starting on the first day of his presidency) asking powerful states for economic and military aid and ordering cutting-edge drones from NATO members – Russia's enemies – so as to fight a successful war? Right intent here is also questionable, in that it does not seem to be the case that Zelenskyy's intent was to spare Ukrainian lives.

The next principle is that of reasonable success. This principle is consequentialist in that the costs and benefits of a campaign must be calculated, and human lives should not be sacrificed in what would be an uneven match. When a nation is threatened by invasion, other forms of retaliation or defence may be available, for example alliances with other small states, so as to equalise the odds, but also diplomacy and compromise.

Is this what Ukraine did? Under what conditions did the Ukrainian president calculate Ukraine's reasonable success? And are there considerations that may lead us to no longer consider Ukraine's war a war of self-defence?

The US-NATO Factor

The argument so far has assumed a clear dichotomy between aggressor and defender, but such simplification does not always correspond to the complexities of security situations that lead to war between states. This section is going to challenge the assumptions that Ukraine is the defender in this war, and that this is a war between two states: Russia and Ukraine.

Where an opponent places its forces is very important in international relations, because expansion and advancement of military forces, increases in influence and partnered operations through new alliances, and establishment of new military bases, can all be perceived as existential threats. Western policies since the end of the Cold War can be seen as attempts by the United States and its European allies to establish a zone around Russia's western border, threatening Russia's security. The United States and NATO have progressively advanced the placement of their military forces towards Russia, all the way to its borders (Estonia, Latvia and Lithuania were accepted in NATO in 2004, in the second post-Cold War enlargement wave). If Russia had taken equivalent action with respect to US territory – say, by placing its military forces in Canada or Mexico – 'Washington would have gone to war and justified that war as a defensive response to the military encroachment of a foreign power'.³⁴

What has been presented as a Western effort to help Ukraine defend itself from Russian aggression can also be interpreted as a campaign to weaken Russia and to degrade its military capacity to fight wars. In March 2022, one month after the start of the war, Chas Freeman, retired US diplomat, commented in an interview:

> Everything we are doing, rather than accelerating an end to the fighting and some compromise, seems to be aimed at prolonging the fighting, assisting the Ukrainian resistance – which is a noble cause, I suppose, but … will result in a lot of dead Ukrainians as well as dead Russians.³⁵

To weaken Russia, a long and bloody war will be required, with many Russian and Ukrainian casualties, but no American ones. In Freeman's words, the United States will feed and prolong this war 'to the last Ukrainian for Ukrainian independence', sparing neither combatants nor civilians. Ukraine will too.

According to the Western narrative, the Russian leader is an expansionist, an evil man, a new Hitler, not motivated by national security concerns, but by aggression similar to that of the Nazis in the Second World War. However, there is a different context in which Russia's violence can be understood. In the past thirty years, since the end of the Cold War, the United States and its European allies have expanded NATO over 1,000 miles eastward, despite assurances that they would not do so; withdrawn from the Antiballistic Missile Treaty and placed antiballistic launch systems that can fire nuclear weapons (such as nuclear-tipped Tomahawk cruise missiles) at Russia in former eastern European and now new NATO states; conducted NATO military exercises near the Russian border; started NATO membership negotiations with Ukraine; armed and trained Ukraine's forces. The US Congressional Research Service reveals that, between 2014 and early 2022, US security assistance to Ukraine amounted to over $4 billion, most of which came from the Department of Defense and the State Department. This amount has increased to $42 billion, following the Russian invasion. This assistance has included logistics support, supplies and services; salaries and stipends; sustainment; weapons replacement; and intelligence support. Through the Joint Multinational Training Group-Ukraine, which was established in 2015, the US Army and military trainers from US allies, provided training, mentoring and doctrinal assistance to the Ukrainian Armed Forces before the war. In 2017, the Trump Administration announced that the United States would provide lethal weapons to Ukraine: 'rocket-propelled grenade launchers, counter-artillery radars, Mark VI patrol boats, electronic warfare detection and secure communications, satellite imagery and analysis capability, counter-unmanned aerial systems (UAS)'.[36]

The anti-ballistic-missile (ABM) system that the United States put into operation in Romania in 2016 uses the Mark-41 'Aegis' missile launchers, which can accommodate Tomahawks with a range of 1,500 miles, capable of striking Moscow and other targets inside Russia. Tomahawks 'can carry hydrogen bomb warheads with selectable yields up to 150 kilotons, roughly ten times that of the atomic bomb that destroyed Hiroshima'.[37] A similar site has been constructed in Poland. John Mearsheimer noted:

> Other NATO countries got in on the act, shipping weapons to Ukraine, training its armed forces and allowing it to participate in joint air and naval exercises. In July 2021, Ukraine and America co-hosted a major naval exercise in the Black Sea region involving navies from 32 countries. Operation Sea Breeze almost provoked Russia to fire at a British naval destroyer that deliberately entered what Russia considers its territorial waters.[38]

In 2020, NATO conducted a live-fire training exercise in Estonia, 70 miles from Russia's border, using missiles with ranges up to 185 miles. The following year, NATO fired twenty-four rockets, simulating an attack on air defence targets. In June 2021, in the meeting of the North Atlantic Council in Brussels, NATO reaffirmed its commitment: 'We reiterate the decision made at the 2008 Bucharest Summit that Ukraine will become a member of the Alliance.'[39] Less than two months before the invasion of Ukraine, on 30 December 2021, Anatoly Antonov, Russian ambassador to the United States, warned that 'everything has its limits', and that if NATO kept 'constructing military-strategic realities imperilling the existence of our country', Russia would be forced 'to create similar vulnerabilities for them. We have come to the point when we have no room to retreat. Military exploration of Ukraine by NATO member states is an existential threat for Russia.'[40] Russia could then also argue that it is fighting a defensive war, due to the existential threat posed by a Western-armed, trained and militarily integrated Ukraine, and by NATO.

Who is responsible for the humanitarian disaster in Ukraine, for the killing and wounding of hundreds of thousands of civilians and soldiers, Ukrainian and Russian? Who bears responsibility for the destruction of homes and the creation of yet another refugee crisis? Who has contributed to the ongoing harm that is being inflicted on the economies of European countries? And who will bear responsibility if the war becomes nuclear?

The obvious answer is Russia, because President Putin declared this war and is directing its conduct. In Suganami's terms, he is responsible for acting with 'belligerent intent'. However, Russia is not the only responsible party, as President Zelenskyy's acts can be seen as threatening and causing the war by provoking the invasion. By actively seeking Western support (financial, material and military) since his election in 2019, he can be said to have been acting 'thoughtlessly', knowing how his actions were perceived by Russia, but not giving serious consideration to the resulting risk of war. It could actually be argued that he has acted 'recklessly', calculating that committing the offensive/threatening acts would lead to more gains, although the probability of war resulting from his acts and its cost would be considerable. Acts carried out after such calculation, risking war in pursuit of long-term gains, are 'reckless' acts. This leads to the conclusion that Ukraine's leader acted thoughtlessly and/or recklessly, and bears responsibility for the war being fought on Ukrainian soil and all its casualties.

On 19 February 2022, just a few days before the Russian invasion, Zelenskyy met with German Chancellor Olaf Scholz in Munich. During this meeting, the German Chancellor offered to broker a peace deal between Ukraine and Russia, telling Zelenskyy:

> Ukraine should renounce its NATO aspirations and declare neutrality as part of a wider European security deal between the West and Russia. The pact would be signed by Mr Putin and Mr Biden, who would jointly guarantee Ukraine's security. Mr Zelensky said Mr Putin couldn't be trusted to uphold such an agreement and that most Ukrainians wanted to join NATO.[41]

Here the issue of probability of success can be addressed, as it relates to Zelenskyy's calculations. If this had been a war between two states, a weak one and a strong one – Russia and Ukraine – similar to a war, say, between the United States and Iraq, then the probability of the weaker state winning the war would be near to zero. We saw how Iraq fell within a few weeks, having had no support from any other state, and having been attacked by a hegemonic power. Ukraine would have probably suffered the same fate, if it had had to defend itself, unsupported, against Russia. What gave the Ukrainian leader confidence in a high probability of victory in a war against Russia was the backing of NATO states. However, what that has enabled is a long war that raises the human cost every day.

If this is not a war between two states, the responsibility of NATO must also taken into account, especially that of the United States, whose acts under two presidents in particular (Trump and Biden) can also be described as 'reckless'. NATO/US assistance has focused on providing capabilities that Ukraine's domestic defence industry cannot produce, increasing its ability to sustain offensive and defensive operations, thus prolonging the war. Provisions have included drones, advanced rocket and missile systems, communication and intelligence support, battle tanks, fighter aircraft and additional air defence capabilities. The provision of security assistance is also focused on improving its 'medium- to long-term capabilities, including transitioning towards more NATO-standard weaponry'.[42] With a large anti-Russian organisation behind it and an aggressive hegemonic state arming it, can it still be argued that Ukraine is fighting a war of self-defence? Can it still be argued that it is fighting a just war, with Russia as the only aggressor and President Putin bearing full responsibility for this war and its horrors? In *How the West Brought War to Ukraine*, Abelow writes that the war between Russia and Ukraine would probably not have taken place, if the United States had not pushed NATO (an alliance created to counter the Russian threat) to the border of Russia; not deployed nuclear-capable missile launch systems in Romania and planned them for Poland; not contributed to the overthrow of the democratically elected Ukrainian government in 2014; not abrogated the ABM treaty and then the intermediate-range nuclear-missile treaty, and then disregarded Russian attempts to negotiate a bilateral moratorium on deployments; not conducted live-fire exercises with rockets in Estonia to practise striking targets inside Russia; not coordinated a thirty-two nation strong military training exercise near Russian territory; finally, not intertwined the US military with that of the Ukraine.

Instead of supporting a negotiated peace in the Donbas region between Kyiv and pro-Russian autonomists, US governments encouraged strong nationalistic forces and poured weapons into Ukraine, which led Ukraine to adopt intransigent positions towards Russia.

The view of this war not as a Russia-Ukraine war, but as a US/NATO-Ukraine-Russia war, results in the following:

- The war in Ukraine may not be a war of self-defence, with Ukraine as the defender.
- The war in Ukraine may not be just.
- The war in Ukraine may involve 'remote' warfare in more ways than one.

Remote warfare then, once again, becomes more than warfare using remote weapon systems, such as drones. This takes us back to America's 3-D wars, that is, wars fought through distance, delegation and darkness. It can then be argued that the war being fought in Ukraine is a US/NATO remote war involving distance and delegation, in other words, a proxy war, or a war of imperialism, which, for just war theory, makes it an unjust war. 'Russia's battle went beyond Ukraine. It was – to a large degree – proxy war against the West', writes Harding.[43] Yet the West, through Ukraine, has also been fighting a proxy war against Russia. In this conflict, both Russia and the West found a new moral purpose that, yet again, has involved militaries and weapons. Does that mean that Russia is 'off the moral hook'? In just war theory, Russia is still the belligerent, the aggressor, so Russia's killings are still immoral and criminal.

From a realist perspective, on the other hand, even in this non-defensive scenario, all killings are justified. That would include all killings and actions taken by the NATO-Ukraine side and those taken by Russia. If 'fear is endemic to states in the international system, and it drives them to compete for power so that they can increase their prospects for survival in a dangerous world',[44] then this (rational) fear and mistrust that drives states to use lethal force, in their struggle for survival, justifies all casualties that occur as part of protecting state interest. From this perspective, all state aggression can be justified and viewed as self-defence. That means that victims like Kateryna Kardash, who 'loved dancing, music and movies', and lost her life in Kyiv on 30 May 2023, in a Russian drone attack, was just collateral damage.[45] Threatening and aggressive behaviour on both sides, especially between two great powers, is part and parcel of the international system. In the context of great power rivalry, Ukraine has made the rational (not moral) choice to ally itself with what it regarded as the stronger party, which once again leads from behind, aiding and arming smaller and weaker states with its own superior technology, to fight its enemy.

From a militarist perspective, there is always a good/moral side and a bad/immoral side, so even if we abandoned the Ukrainian self-defence narrative, and treated this as an American or NATO proxy war, the Ukrainian leader can be seen as having joined the good side, his army fighting with the best weapons, from moral allies, fighting the noble cause of defeating an evil Russia. From a Russian militarist perspective, this war is still being fought for a good cause (to defeat an evil, imperialist West and its pawn – Ukraine), where the virtuous, brave Russian military is protecting the threatened motherland, with great self-sacrifice. Soldiers such as Sgt Nikita Loburets, a squad leader in Russian special forces who died on 20 May 2022, at the age of twenty-one, in a village in eastern Ukraine, are heroes:

> According to an account by his father, Konstantin, Loburets had wanted to be a paratrooper even before he left school in Bryansk, a city about 60 miles (100km) from the border with Ukraine. He began studying martial arts and learned how to parachute-jump before graduation. Eventually, he won a place at the elite Ryazan Higher Airborne School, a training academy for Russian paratroopers, before joining the special forces brigade of the GRU, Russia's military intelligence. Nearly three months into the war, Sgt Loburets and a small unit of Russians were ambushed in a village north of Kharkiv and he was killed, his father said. He was buried in the 'Alley of Heroes' in his home city's cemetery and was posthumously awarded the Order of Courage.[46]

While names of drone operators are never mentioned, Ukrainian father and son Ruslan and Danylo Hulia are remembered:

> Both were professional servicemen. Danylo started service in 2020. In February 2021, his father Ruslan also joined the armed forces. Danylo

Hulia was killed in action on the night of March 10, 2022, near Volnovakha. Ruslan fell in battle of Bakhmut on April 15, 2023.[47]

Both armies can still claim moral superiority over their enemies that have been defined as evil, while the moral army's violence is sanctioned, justified and portrayed as defensive.

Drones' Proliferation and the Global Arms Trade

What the war in Ukraine has highlighted is the breadth of the international arms trade when it comes to the export and supply of lethal weapons in states worldwide. In particular, following the end of the Cold War, weapons dealers from Western states targeted former Eastern bloc countries, as they looked to make a profit on weapons systems.[48] What we also see in the arms trade though is its morally and legally indiscriminate character. Let us take the United Kingdom as an example. The United Kingdom is a leading member of NATO and the UN, a democratic state and human rights proponent. Yet the United Kingdom company Mil-Tec Corporation Ltd supplied $6.5 million of weapons to the Hutu regime in Rwanda, to forces committing genocide, at least from June 1993 to mid-July 1994. Moreover, in the twenty-first century, the military development and production complex is 'sufficiently integrated and powerful to have considerable influence in determining international security policy. It embodies an outlook that prioritises military responses to perceived threats.'[49] The modern-day military-industrial complex exists globally, but especially in powerful states such as the United States, Russia, China, the United Kingdom and France. Each military-industrial complex has interconnected components: arms companies, senior military personnel, civil servants, arms sales, profitable enterprises and a hunt for new enemies, all of which add to a developing culture that can always depend on appeals to patriotism. The whole complex exists as part of a wider security culture, one that privileges national interests and favours military responses to presumed threats. It is a culture of violence, a global culture of war, within which victims are collateral damage and moral values are of little relevance. When it comes to the world's arms dealers, 'their primary function, like that of any other industrial endeavour in a shareholder capitalist system, is to make money for shareholders'.⁵⁰ For arms dealers, a war that escalates, or develops into a violent stalemate, is a perfect war, because in both cases an insatiable demand for arms results, while each side is trying to improve its weaponry and tactics. The profit from such wars is placed above the value of human lives; the value of war is higher than that of peace.

There are several elements in the larger military-industrial complexes in the more powerful states that can help us understand their power and influence, starting with a heavy dependency on arms exports to states that see serious threats to their own security, threats that arms companies are all too ready to emphasise, either in their own marketing processes or by funding appropriately orientated research. States with thriving arms industries will provide diplomatic and other support for arms sales, mount exhibitions, facilitate travel and provide intelligence. The complex must demonstrate 'need', so that if an arms company sees one of its weapons systems used in a war it will publicise that through the industry and in the defence and security journals, not infrequently under the heading 'combat proven'.

Many arms-industry leaders may choose to view themselves as patriotic guardians of their country, or as guardians of global security, but the system in which they operate raises grave ethical questions. As Ukraine continues to receive weapons and ammunition from NATO members, the war is becoming long, perfect for arms companies to profit from.

According to the Stockholm International Peace Research Institute (SIPRI), the United States is the world's largest arms exporter responsible for 38.6 per cent of international arms sales between 2017 and 2021, up from 32.2 per cent between 2012 and 2016. It has supplied arms to more than 100 countries, in most of South America, Africa, the Middle East, Europe and Australia.⁵¹ The world's leading exporter of combat drones, however, is China. From Saudi Arabia to Myanmar, from Iraq to Ethiopia, Rasheed writes:

> governments and militaries across the globe are stockpiling Chinese combat drones and deploying them on the battlefield. In Yemen, a Saudi-led coalition has dispatched the Chinese aircraft, also known as uncrewed aerial vehicles or UAVs, as part of a devastating air campaign that has killed more than 8,000 Yemeni civilians in the past eight years.

In Iraq, authorities say they used Chinese drones to carry out more than 260 air raids against ISIL (ISIS) targets as of mid-2018, with a success rate of nearly 100 percent.[52]

The military in Myanmar has used Chinese drones to conduct hundreds of air attacks on civilians and ethnic armed groups that are opposed to its power grab through a military coup in 2021, increasing its airstrikes against civilian locations by 141 per cent in 2022. In Ethiopia, the prime minister's fleet of imported Chinese, Iranian and Turkish drones was critical in helping his forces thwart a rebel march in 2021. Other buyers of China's combat drones – aircraft that can fire air-to-surface missiles – include Morocco, Egypt, Algeria, the United Arab Emirates, Pakistan and Serbia. Data from SIPRI shows that China has delivered 282 combat drones to seventeen countries in the past few years, making it the world's leading exporter of the weaponised aircraft.

The focus of China's drone programme has been on replicating the capabilities of other countries – surveillance, attack and electronic warfare capabilities – according to Akhil Kadidal, an aviation reporter at Janes, a media outlet specialising in defence issues. China's bestselling drone, similar to the US-made MQ-9 Reaper, is the Caihong 4, while the popular Wing Loong 2 is similar to the US-built MQ-1 Predator. China's UAV programmes suggest that Beijing is interested in creating a better platform than their Western counterparts, and the Wing Loong 2 and 3 are examples of this. The Chinese claim that both these UAVs are faster than their US counterparts and capable of carrying a greater weapons payload. The Chinese drones are also much cheaper, making them more attractive to global buyers. For instance, the CH-4 and the Wing Loong 2 are estimated to cost between $1 million and $2 million, while the Reaper costs $16 million and the Predator $4 million, according to the Center for Strategic and International Studies, the US-based think tank. The cheaper price tag means that governments can buy the drones in larger quantities and deploy them as they see fit, even in violation of human rights and international law, and irrespective of any moral considerations.

Many African nations are beginning to accumulate UAVs. Niger and Nigeria, for example, have reportedly acquired armed drones to patrol their vast hinterlands for jihadist groups. But often, 'no one knows who has what until they are seen in action because of the opaqueness of the arms trade'.[53] There are few safeguards, policy frameworks, or codes of conduct for the new technology, which results in human rights abuses and the terrorising of civilians.

Writing about Israel, Feinstein observes that unmanned, armed observation points identify and open fire on whoever comes close to the barrier. He argues that this is the creation of 'a robotic warfare that accelerates and intensifies the process of dehumanization and non-culpability for death – the very factors that have enabled mass killings and genocide to occur from Auschwitz to Kigali'.[54]

During the twentieth century, the trade in arms made viable conflicts that led to the killing of 231 million people. The first two decades of the twenty-first century have already seen short, long and perpetual wars, conventional, irregular, remote and proxy, at significant – and rising – human cost. Rather than committing to universal human rights, equality and justice, the pattern continues to be the production of more deadly weapons, the unscrupulous taking of life and the profiting from the suffering of others. While for the realist it is interests, power and national security that drive and should drive all conflict, with little regard for morality and loss of life, and for the militarist it is right that there should be military responses to presumed threats, from a just war perspective what is problematic and unethical is the resistance of such approaches to peaceful solutions, the increasing production, indiscriminate exporting and use of new war technologies, as well as the lack of accountability, when human lives are taken. And as war is becoming more automated and robotic, there are more and more opportunities to lose control of the violence and the agonies, physical and metal, it unleashes.

Conclusion: A World of Drones

This chapter began by examining the ethics surrounding Ukraine's war of national defence. It first addressed issues around claims of self-defence and then explored the remote weapons used, in a war that has provided a testing ground for foreign drone powers: the Turkish-manufactured TB2 Bayraktar, the Kamikaze Switchblade and the ALTIUS-600 drones from the United States, the UK-provided Storm Shadow long-range missiles, the American-made Phoenix Ghost and Polish-developed Warmate drones, drone boats, the R18, developed by the Ukrainian Aerorozvidka, as well as 'dronations' from civilians. Russia has used Lancet-3M or Product 52 and has also been employing Iranian-supplied Shahed-136 kamikaze drones, renamed Geranium-2.

Ukraine's drone wars were assessed through a realist justification and a militarist critique. Ultimately, the use of remote warfare by Ukrainian forces was justified not only by realism, but also by militarism, as part of a rational (realism) and morally justified (militarism) wider military response to a threat. While just war would justify a war of self-defence, there is still the question of leader responsibility to avoid placing citizens at risk of death. In Ukraine's case, the leader, President Zelenskyy, had other options available, options that would have avoided taking Ukrainian citizens to war.

The US-NATO factor in this war further highlighted the responsibility of Ukrainian leaders, as well as challenged the argument that this was a war of self-defence on the part of Ukraine, whose actions over two decades – thoughtless and reckless – could be seen as causes of war. So keen was Ukraine to join NATO, a military alliance created precisely to counter the Soviets and Russia, that in 2003 it sent forces to the newly invaded and occupied Iraq, to fight alongside the invading/occupying forces: the United States and the United Kingdom. According to Iraq Body Count data collected in 2004, Ukrainian troops killed thirteen Iraqi civilians as they shot at protesters in January and April. A speech made by Russian President Putin in Munich, in February 2007, is very revealing of the perceived Western threat by Russia:

> One state and, of course, first and foremost the US, has overstepped its national borders in every way. This is visible in the economic, political, cultural and educational policies it imposes on other nations ... I think it is obvious that NATO expansion does not have any relation with the modernisation of the alliance itself or with ensuring security in Europe. On the contrary, it represents a serious provocation that reduces the level of mutual trust. And we have the right to ask: against whom is this expansion intended? And what happened to the assurances our Western partners made after the dissolution of the Warsaw Pact?[55]

Putin concludes that, unlike the UN, NATO is not a universal organisation, but a military and political alliance. Led by the United States, NATO has launched sustained military operations in the Balkans, Libya and Afghanistan. As for the United States, it has invaded and intervened militarily in several more countries, such as Iraq, Pakistan, Somalia and Syria. No matter how often it is claimed that NATO is purely a defensive alliance, its vast offensive military power projection capabilities that have been used in past wars tell a different story. NATO expansion is in line with US interests, it is not dictated or influenced by ethics, altruism or humanitarian considerations. Ukraine's actions are similarly in line with its own strategic interests and not guided by ethics, as the case of Iraq showed. If Ukraine joined the alliance, NATO's soldiers (Ukrainian or otherwise) would be stationed in Kharkiv, in Luhansk, in Donetsk, in Mariupol.

The US-NATO-Ukraine factor in this war calls into question the self-defence argument and therefore the *jus ad bellum* claim of Ukraine fighting a just war. If the war is just, then Ukraine's drone warfare – especially as it only kills soldiers – can be called just and ethical. However, if the war is not defensive or just, then the ethics regarding all its killings are dubious at best. Moreover, the involvement of NATO members, especially the United States and the United Kingdom, brings us back to the ethics of the War on Terror, where great/hegemonic powers fight wars of imperialism guided by self-interest, profit and the desire for power. Just like in the War on Terror, in Ukraine's proxy war we once again have the 3-D war of remoteness: distance, delegation and darkness.

Most worryingly of all, what this war has made clear is the proliferation of drones in the utterly immoral and profit-driven global arms trade. With American, Iranian, British and Turkish-made weapons spreading death and destruction across Ukraine, the world is reminded of the dangers of UAS. From 2021 to 2022, drone sales went up 57 per cent, an increase that has made it difficult 'to effectively ensure every echelon down to the operators of counter-drone systems is on the same page when it comes to strategic vision, operational mission, and tactical employment',[56] it has also made it difficult to keep up with the legal and moral implications, in this fast-paced evolution of warfare.

More than ten states have so far conducted drone strikes: the United States, Israel, the United Kingdom, Pakistan, Iraq, Nigeria, Iran, Turkey, Azerbaijan, Russia and the United Arab Emirates, while many other countries, including Saudi Arabia, India and China, maintain armed drones in their arsenals. All these states are navigating their own rules in the air, as there does not exist an international agreed-upon set of regulations. In the last few years, Serbia, Germany, Indonesia, Singapore, Algeria, Qatar and the Netherlands have acquired armed drones.

Serbia

In September 2018, Serbia announced its procurement of unmanned combat aerial vehicles from China. The model that was chosen by the Serbian government was the Wing Loong I UAV.

Germany

In June 2018, Germany's Bundestag Budget Committee approved the lease of armed drones from Israel.

Indonesia

At the end of 2019, Indonesia unveiled to the public its first MALE UAV called the Elang Hitam (or Black Eagle) on 30 December.

Singapore

The Republic of Singapore Air Force, in a now deleted image, showed the existence of an Israel Aerospace Industries Heron 1 MALE UAV in October 2019.

Algeria

Algeria tested the Chinese CH-4 UCAV in 2014. It now operates CH-3 and CH-4 UAVs. It also has UAVs from the United Arab Emirates, the Adcom Yabhon United 40, which is named locally as the Algeria 54, and the Yabhon Flash-20.

Qatar

The Qatar Armed Forces and the Turkish aircraft manufacturer 'Baykar Machinery' made a deal including six 'Bayraktar TB2' military drones and control stations, as well as technical and logistical equipment for these aircraft.

Canada

In late 2019, the government of Canada approved requests for proposals to include new armed UAVs into its military. Both the Heron and MQ-9 are likely to be part of the fleet. Canada has long sought an armed drone fleet, especially after NATO intervention in Libya.

Netherlands

In March 2019, the US Department of Defense released a statement that the Netherlands was purchasing four MQ-9 Reaper UAVs.

Israel

Israel exports to fifty-six states: Ukraine, France, Thailand, Russia, Indonesia, Germany, Spain, Netherlands, Kazakhstan, Colombia, China, Australia, Vietnam, the United States, Turkmenistan, Sweden, South Korea, Finland, Canada, Zambia, the United Kingdom, Uganda, Turkey, Switzerland, Sri Lanka, Singapore, Serbia, Poland, Philippines, Nigeria, Mexico, Kenya, Hungary, Georgia, Ethiopia, Dominican Republic, Cyprus, Chile, Cameroon, Brazil, Belgium, Taiwan, South Africa, Slovakia, Serbia, Peru, Ivory Coast, Ireland, India, Iceland, Honduras, Ecuador, Croatia, Botswana, Azerbaijan and Angola.

The United States

The United States exports to fifty-five states: Ukraine, France, Denmark, Thailand, Indonesia, Germany, the United Arab Emirates, Spain, Pakistan, Netherlands, Egypt, Colombia, Australia, Vietnam, Sweden, South Korea, New Zealand, Malaysia, Italy, Canada, Uzbekistan, the United Kingdom, Uganda, Turkey, Syria, Switzerland, Singapore, Qatar, Portugal, Poland, Philippines, Mexico, Lithuania, Lebanon, Kenya, Israel, Iraq, Hungary, Cameroon, Belgium, Austria, Tunisia, Romania, Panama, Oman, Norway, North Macedonia, Luxembourg, Latvia, Japan, Estonia, Czech Republic, Burundi, Bulgaria and Afghanistan.

China

Interestingly, China also exports armed drones to some of the same states: Ukraine, France, Denmark, Russia, Indonesia, Germany, the United Arab Emirates, Egypt, Colombia, Australia, Kazakhstan, Turkmenistan, Finland, Uzbekistan, Zambia, New Zealand, Sri Lanka, Serbia, Portugal, Nigeria, Pakistan, Lithuania, Israel, Iraq, Ethiopia, Dominican Republic, Chile and Brazil. Drone exporters to other states include Austria, France, Canada, Germany, Iran, Estonia and Turkey.[57]

That's as far as states, but non-state actors all over the world are also using armed drones. Palestinian Islamic Jihad, for example, has used drones against Israeli forces. Venezuelan military defectors attacked President Maduro with drones armed with explosives, during a speech commemorating the Bolivarian National Guard's eighty-first anniversary at a military parade in 2018. In Nigeria, Boko Haram fighters now 'have more sophisticated drones than the military'.[58] The militant group Harakat Tahrir al-Sham, with ties to al-Qaeda, attacked a Russian military base in 2018 using a 'swarm' of armed drones rigged with explosives. The Kurdish terrorist group PKK carries out drone strikes against Turkish soldiers, while the Libyan National Army has conducted drone strikes against Libyan government forces since 2017, using Chinese-made Wing Loong II drones supplied by the United Arab Emirates.

In Syria, at least four groups have used drones: Faylaq al-Sham (a Sunni rebel group fighting pro-Assad forces), the Turkistan Islamic Party (allied with al-Qaeda), Saraya al-Khorasani (Iranian-backed Iraqi Shia militia fighting ISIS forces) and Suqour al-Sham Brigades (part of the National Liberation Front). In the Philippines, the pro-ISIS Maute militants used drones against government forces during the Marawi battle. In Mexico in 2017 an armed drone was discovered in the arsenal of Cártel de Jalisco Nueva Generación, an organised crime syndicate. ISIS has trafficked commercial drone technology from at least sixteen companies across at least seven countries, attaching munitions to quadcopters and small fixed-wing drones. The kamikaze quadcopter drones had munitions strapped to them that could independently release.[59] In eastern Ukraine, the separatist group Donetsk People's Republic has used armed drones supplied by Russia since 2015.

Drones can be launched in massed and swarm attacks. The first kind 'resembles several birds with decentralized flight patterns picking and choosing different prey while the second resembles an organized flock of birds converging on a single target'.[60] Swarm attacks are particularly lethal, as they feature coordinated command and control, where a single operator uses an algorithm or tactical operations centre. Multiple drones can communicate with each other remotely, and like a swarm of bees, they form a deadly and autonomous aerial army ripe for accidents. States are more likely to practise such attacks, which involve large numbers of drones, and China and Russia have focused on developing or using small unmanned aerial systems used in swarm attacks. In the Middle East, there have been Houthi drone swarm attacks in Saudi Arabia, which has accused Iran of arming the Houthis. On 11 March 2022, Yemen's Houthi group claimed a drone attack on a refinery in the Saudi capital Riyadh. The Iran-aligned Houthi movement targeted a Saudi Aramco refinery using three Samad-3 drones, while six Samad-1 drones were also fired at Aramco facilities in the Saudi cities of Jizan and Abha.[61]

The war in Ukraine has highlighted a growing global arms trade that involves weapons with as yet unclear ethics, legality and questions around accountability, in a global war business where states profit from the development and sale of weapons to a variety of other states and to non-state actors, which use them for their own purposes. At the same time, the exporting states also use this technology, either domestically or abroad, sometimes in the client states. The example of the United States and Saudi Arabia shows this problem of moral and legal accountability. In 2014, the Saudi military ordered two CH-4 and five Wing Loong armed drones from China. By 2016, two Wing Loong drones had been deployed to the Najran Province, near the border with Yemen, where soon one of them was seen downed inside Yemen. The following year Saudi Arabia ordered 300 Wing Loongs from China and it was reported that a manufacturing plant would be opened in Saudi Arabia to facilitate this sale:

> Saudi Arabia has used its armed drones to carry out strikes in Yemen … In addition, US drones based in Saudi Arabia may be among those launching strikes into Yemen, further complicating attribution. Working out which air force has carried out what strikes is one of the inherent

problems of drone technology and is likely to cause further accountability problems in the future.[62]

The arrival of artificial intelligence and the development of sophisticated UAVs have made drones more lethal and even more difficult to track, to investigate and to legislate. Drone attacks can come out of the blue, with no restrictions or consequences, because there are no rules guiding drone transfers, exports, imports and usage. Meanwhile, drone technology continues to evolve and, after the current war in Ukraine is over, what will likely follow is grey zone conflict: distant, proxy, dark, profit and interest-driven, and through delegation. And war, more than ever before, becomes an amoral or immoral business, most incompatible with just war principles.

Ukraine has 'tragically become a battle lab' for war technology, according to Ben Wallace, Secretary of State of the United Kingdom, but hopefully, the lessons learnt from it will inform the future. 'New technologies are not gimmicks, they're fundamentally key to how we fight a modern war', he declared a year-and-a-half into the war, adding that analysing the strategies playing out in Ukraine would help make sure that the United Kingdom would be 'match fit for any future conflict'.[63] One of the lessons learnt is the power and use of electronic warfare, which is going 'up the priority list'. Once again, it is the realist's strategy, rather than any moral considerations, that takes priority and drives the action, the aim of which is to win at any cost, using the latest weapons with little to no control or regulation and with no accountability.

Conclusion: Remote Warfare and the (New) Ethics of War

A clear majority of states see the need to ensure meaningful human control over the use of force. It's time now for them to lead in order to prevent the catastrophic humanitarian consequences of killer robots.[1]

For the first time ever, in 2021, the majority of the 125 nations that belong to the UN Convention on Certain Conventional Weapons asked for new laws to be introduced on killer robots. Yet in December 2021, the UN failed to agree on banning the use and development of so-called 'slaughterbots', raising alarm bells among experts in artificial intelligence, military strategy, disarmament and humanitarian law. Slaughterbots are weapons that select and apply force to targets without human intervention, making their decisions with AI software, which is essentially a series of algorithms.[2] States developing these weapons – including the United States, the United Kingdom and Russia – were in opposition, which made a unilateral agreement impossible. The conference concluded with the group pledging to intensify discussions and consider possible steps that would be acceptable to all.

Slaughterbots were, at that time, already being used in Libya. Turkish-made Kargu drones, small portable rotary wing attack drones that are supposed to provide precision strike capabilities, were used in Libya's civil war, according to a UN Security Council report published in March 2021. According to the report, Turkey both manufactured and used remote weapons in Libya: 'Their fixed-wing fighter ground attack aircraft, rotary-wing attack helicopters (Mi24/35) and unmanned combat aerial vehicle (Wing Loong II) provided local air superiority over the majority of the country.'[3] The Government of National Accord Affiliated Forces (GNA-AF) was supported with self-propelled guns and T-122 Sakarya multi-launch rocket systems firing extended range precision munitions against the battle tanks and artillery used by the

Haftar Affiliated Forces (HAF). As the HAF were retreating, they were hunted down and remotely engaged by the unmanned combat aerial vehicles or the lethal autonomous weapons systems: the Kargu-2 and other loitering munitions, which were programmed to attack:

> without requiring data connectivity between the operator and the munition: in effect, a true 'fire, forget and find' capability ... The concentrated firepower and situational awareness that those new battlefield technologies provided was a significant force multiplier for the ground units of GNA-AF, which slowly degraded the HAF operational capability. The latter's units were neither trained nor motivated to defend against the effective use of this new technology and usually retreated in disarray. Once in retreat, they were subject to continual harassment from the unmanned combat aerial vehicles and lethal autonomous weapons systems.[4]

The report celebrated the use of UAVs and credited them with the successful defeat of the HAF by the officially endorsed government forces. Since then, the development and use of killer robots has only accelerated, as they are easy and cheap to mass produce. They are being used both by states and non-state actors, including terrorists. The companies that make the killer robots have been developing AI systems that can identify the thermal signature of a human target, or identify their face via a camera, this way distinguishing between combatants and non-combatants, with accuracy and precision – or so they claim.

But the International Committee of the Red Cross, viewed by many as the custodian of the laws of war, has called for the prohibition of autonomous weapons that are designed and/or used to target human beings. Richard Moyes, coordinator of the Stop Killer Robots campaign, has argued that government leaders need to draw a moral and legal line for humanity against the killing of people by machines. Emilia Javorsky, leader of the Future of Life Institute's advocacy programme on autonomous weapons, warned that after the 'epic failure' of the UN Convention on Certain Conventional Weapons to agree on anything concrete, it is clear that this forum (whose unanimity requirement makes it easy for any state with a vested interest to derail it) 'is utterly incapable of taking seriously, let alone meaningfully addressing, the urgent threats posed by emerging technologies such as artificial intelligence'.

Verity Coyle, senior adviser at Amnesty International, also warned that the window of opportunity to regulate killer robots is growing ever smaller: 'The CCW [Convention on Certain Conventional Weapons] has once again demonstrated its inability to make meaningful progress – it's now time that committed states take the lead on an external process that can deliver the type of breakthrough we've previously seen on landmines and cluster munitions.'[5] Lethal autonomous weapons can be compared to bioweapons, in terms of control; it can be argued that bioweapons are also autonomous weapons in the sense that, once released, they can attack their targets without the need for human intervention.

The use of drones in global conflicts and wars has already featured mass, swarm, lethal attacks, with the command side being limited to what James Rogers calls 'human-in-the-loop' control. Rogers explores three types of human involvement and command when it comes to armed drones: in the loop, on the loop and outside the loop. Currently, humans are in the loop: they are operating the weapons remotely. On the loop means that humans are not in direct control at all times but take control over any decisions that the machines make. Of most concern, regarding future conflicts, would be when the human is outside the loop of control. In the near future, it is possible that AI-powered drones will execute entire missions without human intervention, reporting the result after completion of the attack. A human should always be in the loop of control when the fate of another human being is at risk – these decisions should never be left to a computer. It seems inevitable now that drones will play a pivotal role in future conflicts and they will be used by state forces and by non-state actors, which now have the capacity to manufacture, deploy and even supply weaponised drones to other militant groups. The twenty-first century appears to be an age of unchecked and uncontrollable drone proliferation, both to states and to non-state actors all over the world. Rogers predicts that fully autonomous systems will be part of the state and the non-state arsenal by 2040, with implications regarding the regulation of these weapons, if they can even be regulated.

Just war principles like distinction and proportionality cannot be programmed into a computer, which means that drones will never be able to abide by international humanitarian law. Whether used by state or by non-state actors, a weapon that is less than accurate and violates international humanitarian law might even be considered better – more terrorful: 'AI-powered weapons will be weapons of terror and weapons of error.'[6] They will also be weapons that may be unattributable, for in many cases it will be very difficult to determine who was behind the AI-powered weapons that killed human beings, and – if more than one party was behind them – which one, if any, to hold accountable.

Modern militaries are deploying autonomous weapons systems to survey and identify potential threats (on and off the battlefield), to improve the speed with which they respond to threats, and to eliminate those threats – that is, to kill. A human operator will set the mission parameters, but then an autonomous system can take over and execute the mission. Sauer argues that while AI systems are very fast at categorising objects, they lack the nuance required to make life-and-death decisions: 'Machines don't understand anything, they're just good at matching patterns.'[7] When deciding how much power an autonomous system has, the impacts regarding international humanitarian law and ethics need to be considered, because allowing artificial intelligence unregulated control could be a runaway nightmare. Risks include unpredictable outcomes, swift and unwanted escalations from crises to wars, or higher levels of violence. Even if armed drones were made to distinguish between combatants and non-combatants, in accordance with just war principles, that would still not make the killings just or ethical:

> The narrow focus on discrimination implies that, as long as civilians remain unharmed, algorithmically attacking combatants could be acceptable. But combatants, too, are imbued with human dignity – and being killed by a mindless machine that is not a moral agent is infringing on that dignity.[8]

The use of armed drones must have limits and constraints in all areas: in curbing ethical risks, in controlling the flow of arms, and in safeguarding international humanitarian law. However, international humanitarian law permits many acts that can and do cause incidental harm, and new military technologies create further accountability gaps. Rules grounded on just war's moral-legal principles; for instance, forbid the targeting of civilians, yet an attack that incidentally results in civilian harm – even foreseeable and extensive civilian harm – may be deemed lawful. If the attack was (reasonably) expected to lead to collateral damage that was not excessive, compared to the anticipated military benefits, the proportionality requirement is met and the operation is proportional, regardless of the civilian losses. This unease that exists in the just war approach is exacerbated by new weapons like drones, which exploit the grey zone of incidental harm.

Moreover, while 'precise' weapons are celebrated for minimising risks to combatants and to civilians, they also create a precision paradox, because the ability to engage in more 'precise' strikes means that a military can lawfully engage in more strikes, because 'an operation that might have once risked too much civilian harm and thus failed the proportionality requirement may be rendered newly lawful'.[9] So, there are more attacks and, consequently, more permitted civilian harm, either as anticipated collateral damage or unforeseen accidental harm. The new technology is not creating a new legal-ethical problem, it is expanding an existing one: the problem of unintended civilian harm. It also highlights the fact that there is no international accountability mechanism for most unintended civilian harms in war. Under international law, no one is accountable for the harmful consequences of lawful acts in armed conflict, and those harmful consequences all too frequently involve innocent civilians. In failing to establish a deterrent, the law, with its just war ethics, arguably facilitates unintended civilian harm.[10] New technologies perpetuate and expand this harm to civilians.

Civilian harm in armed conflicts has been documented by casualty recorders worldwide, and data collected was cited in Chapters 4 and 5 of this volume: Iraq's casualties have been recorded and documented in an online database by Iraq Body Count, for example, while casualties in Ukraine are being recorded by the UN and Memorial, among others. Casualty recording is the systematic collection of information on individual deaths and/or injuries, in situations of violence and of armed conflict. The data collected by casualty recorders at the very least indicates that moral or legal violations of human rights have occurred, which makes casualty recorders invaluable for providing the evidence on which moral and legal assessments can be made, when it comes to the impact of any weapon on humans.

In its 2023 annual report of the UN High Commissioner for Human Rights and reports of the Office of the High Commissioner and the Secretary-General Promotion and protection of all human rights, the UN recognises the importance of casualty recording, for insight and analysis into critical aspects of armed conflicts and situations of violence:

> Casualty recording has an impact on protection, compliance with international law, early warning, prevention, accountability, access to services and reparations, among others. Through its multiplicity of contexts, actors and approaches, casualty recording can become an integral part of responses to violence and conflict. The work should be further supported to assist in stopping or mitigating harm to civilians and victims to ensure that all casualties are identified.[11]

If we want to protect civilians under threat from violence and suffering, if we want to achieve peace, if we want to address civilian harm, and if we want to understand war, we must be prepared to see it in all its ugliness, in all the pain and devastation it brings. Unfortunately, states will often deny the harms caused by their actions, refuse to acknowledge civilian harm, which is an easy response when civilians have been killed remotely, as the Al Badiya school strike shows.

In Al Mansoura (Syria), the abandoned Al Badiya school (a large, three-storey building) housed refugees and internally displaced persons that had fled violence elsewhere in the country. Al Mansoura was in 2017 part of an ISIS-controlled area. By March 2017, 200–400 people were estimated to be living in the abandoned school. On the night of 20–21 March 2017, the building was deliberately struck by the US-led coalition against ISIS, destroying the building and killing almost everyone inside. The coalition, however, denied that the strike had killed civilians:

> We had multiple corroborating intelligence sources from various types of intelligence that told us the enemy was using that school. And we observed it. And we saw what we expected to see. We struck it. We saw what we expected to see. Afterwards, we got an allegation that it wasn't ISIS fighters in there ... it was instead refugees of some sort in the school. Yet, not seeing any corroborating evidence of that. In fact, everything we've seen since then suggests that it was the 30 or so ISIS fighters that we expected to be there.[12]

While the US-led coalition denied any civilian harm, Baladi News put the number killed at 200 civilians – mostly women and children – with dozens more injured.[13] As many as 420 people were reported killed by Mansoura in its People's Eyes.[14] It was only in June 2018 that the coalition admitted to killing at least forty civilians (among them sixteen children). As a consequence of denials and underreporting by states, (incorrect) precision claims are made and the true magnitude of wrongful killings is mostly unknown. It is only when organisations like Iraq Body Count, Airwars, the Syrian Observatory for Human Rights, Amnesty International and Human Rights Watch monitor and challenge, and do so with full commitment and dedication to human rights, ethics and justice (in the face of great adversity by powerful states), that some of the truth becomes known. In the case of Al-Badiya school, the coalition only reopened the case 'after a field study by Human Rights Watch made their continued denial of civilian deaths impossible. This is not an isolated incident, but part of a larger problem.'[15] Unobservable civilian harm in remote strikes is the main reason why civilian harm continues to be denied or undercounted. In the words of Air Marshall Bagwell, in an interview with Drone Wars UK, 'We cannot see through rubble'.[16]

Mark Lattimer, Director of the Ceasefire Centre for Civilian Rights, points out that while in the past the assumption was that, once war broke out, laws of human rights would give way to the laws of armed conflict (international humanitarian law), international courts have confirmed that human rights law continues to apply during armed conflict, although it needs to be interpreted in a way that takes account of international humanitarian law. However, the further complication that arises in drone warfare is the difficulty in determining whether a particular situation can be classified as an armed conflict or a war. A soldier/combatant can lawfully kill other combatants in war, but in cases of drone killings it is sometimes difficult to determine if an armed conflict exists or whether we are dealing with a targeted assassination. How you classify a situation where armed drones are killing people has an impact on whether particular laws apply – and if so, which laws. Simply put, the laws of armed conflict apply only if there is an armed conflict. A targeted killing outside of an armed conflict is an arbitrary or extra-judicial killing: effectively, an assassination.[17] In war there are always clashing ethics, but that is not a good reason to abandon discussions on ethics; in fact, ethical issues need to take centre stage.

An ethical approach to war requires that we do not consider only the lives and wellbeing of 'our' people, as realism asserts. An ethical approach requires that we do not enforce our idea of justice and of right with deadly weapons, as militarism demands our moral warriors do. An ethical approach requires force used only in self-defence, based on the principles of just war, but is there a purely defensive war? Even if there is defence involved, does the defence of anything justify the taking of human life? And what is war? These days all manner of violent actions and policies are given the name 'war', such as the so-called 'war on drugs'. Is this a war, in the sense of the Russia-Ukraine war, or is it the targeting of individuals defined as threats to something? It is worth remembering here that targeted assassinations have, for decades, been conducted by governments in the 'war on drugs', against drug lords, which means that the victims do not even need to be combatants, to be regarded as legitimate targets in remote wars.

For Homer the paradigm of human excellence was the warrior. In his poems, a virtue is a quality, the manifestation of which enables someone to do what their social role requires. The primary role is that of the warrior king, and the key virtues are those that enable a man to excel in combat.[18] In the Homeric model we see militarism and realism meet: the good or virtuous life for a man is in fulfilling his social role, so his goodness lies in his utility to his society or state, of which excelling in combat is the greatest. In war, this model easily combines the two approaches, allowing for the warriors to be the noblest and most virtuous of citizens, while at the same time prioritising the wellbeing of those charged with the protection of the state. For the realist, remote warfare is the best way to protect those who protect the state – those who have the highest social role. Militarists, however, concerned about losing the heroic elements of war in remote warfare, look at drones with little enthusiasm. But is war heroic? There may be heroism in risking or giving one's own life, but there is nothing heroic in extinguishing another life. There are no moral armies.

What any moral understanding of war needs is less concern with power and more concern with humility and humanity. Morality requires that security is understood not at the level of states and their armies, nor at the level of militant groups and their 'right fighters', but at the level of the ordinary person and their needs. It demands that, as Bernard Williams argued, each person is owed an effort at identification, each person should not be regarded as the surface to which a certain label can be applied,[19] because each life is unique, irreplaceable and precious. Richard Rorty reminds us that the bad people are no less rational, no more prejudiced, than the good people who respect otherness.[20] That necessarily means that, if we are concerned with ethics, killing a person must made more difficult and less frequent, not easier and commonplace.

As this volume is being completed, UN Secretary-General Antonio Guterres is calling for new international law to regulate and prohibit killer robots. On 20 July 2023, Guterres released 'A New Agenda for Peace' issuing an urgent call to states to adopt a legal treaty to prohibit and regulate autonomous weapons systems by 2026. This constitutes an important recognition that new technologies pose fundamental humanitarian, legal, security and ethical concerns that directly threaten human rights and freedoms. According to the New Agenda for Peace, member states should conclude by 2026 'a legally binding instrument to prohibit lethal autonomous weapon systems that function without human control or oversight, and which cannot be used in compliance with international humanitarian law, and to regulate all other types of autonomous weapons systems',[21] as there is a moral need to safeguard against the risks that such weapons pose to humanity. The agenda warns that the design, development and use of these systems will continue to raise humanitarian, legal, security and ethical concerns, recognising the perils of weaponising new technologies and their potential to revolutionise conflict dynamics. One of twelve Recommendations for Action considers the transformative potential of emerging technologies in conflict and warfare, and the risks posed to human rights due to issues with accuracy, reliability, human control and algorithmic bias.

Operation Guardian of the Walls saw 4,300 rockets and drones fired towards Tel Aviv and central and southern Israel in 2021, with 20 per cent of the strikes breaking through the world-leading Iron Dome defences that Israeli officials had hailed as a 'game-changer' that would save civilian lives. Yet even the most advanced air defence system can be overwhelmed by the current generation of drone-assisted saturation strikes. Much like in the Second World War, where 'the bomber would always get through', modern killer drones will always get through too. The result is death.

Remote killing makes it easy on the killers, in the death chambers, under the skies of Germany, Britain, Japan, Iraq and Ukraine, where the air – a source of life – became a lethal space for millions. The states' only concern was that their own security agents were protected – physically, mentally and morally – as each of those millions was drawing a last breath. What kind of ethics are concerned only or primarily with the wellbeing of the killers? A state, a society, a world that does not concern itself with the act of killing – that is concerned only that the hearts and minds of combatants are not burdened by the accompanying responsibility – has lost touch with fundamental humanitarian norms. The more we distance ourselves from killing, the easier killing becomes, but morality requires that we make it harder, not easier, to take a life.

In some chapters of this book some victims are named, in order to humanise, to dignify, to memorialise the dead, and also to bring home the enormity of the loss that only names and details about identity – rather than numbers – can bring. Once again, let us remember human losses. It is apt that a book on ethics concludes with the real, the most heartbreaking and morally unacceptable losses of any war:

- Anne Frank, German-born Jewish schoolgirl, killed remotely at sixteen in a death camp in 1945.
- Ludwig Czerny, German film director, killed remotely at fifty-six, as he helped a woman carry her pushchair down to an air-raid shelter, in 1941.
- Isobel Thornley, British historian, killed remotely in her home at forty-eight, in 1941.
- Branimir Stanijanovic, Serbian boy, killed remotely at the age of six with his parents Dima and Vidosav, while fleeing in a train, in 1999.
- Amina Ayman Ahmed, Iraqi girl, killed remotely at the age of three, in 2017.
- Yuliia Zubchenko, Ukrainian medic, killed remotely at twenty-seven, in 2022.

Notes

Introduction

1. Friedrich Nietzsche, 'Thus Spake Zarathustra', *www.naturalthinker.net/dquinn/Zarathustra/Zarathustra02.html* (accessed 15 August 2023).
2. Michael Walzer, *Just and Unjust Wars*, 3rd edn (New York: Basic Books, 2000).
3. James Eastwood, *Ethics as a Weapon of War: Militarism and Morality in Israel* (Cambridge: Cambridge University Press, 2020).
4. Shaun Best, 'Zygmunt Bauman: On what it means to be included', *Power and Education*, 8/2 (2016), 124–39, *https://journals.sagepub.com/doi/full/10.1177/1757743816649197* (accessed 12 June 2024).
5. Lawrence Freedman, *The Future of War* (London: Penguin Random House, 2017).
6. Brennan Deveraux, 'Loitering Munitions in Ukraine and Beyond', *Texas National Security Review* (2022), *https://warontherocks.com/2022/04/loitering-munitions-in-ukraine-and-beyond/* (accessed 15 August 2023).
7. James Der Derian, *Virtuous War*, 2nd edn (New York and London: Routledge, 2009), p. xxxi.
8. Der Derian, *Virtuous War*, p. xxxi.
9. Der Derian, *Virtuous War*, p. xxxii.
10. Lorraine Dowler and Joanne Sharp, 'A Feminist Geopolitics?', *Space and Polity*, 5/3 (2001), 168.
11. Klaus Dodds, *Global Geopolitics: A Critical Introduction* (London: Pearson Education, 2005), p. 228.
12. Rebecca Crootof, 'AI and the Actual IHL Accountability Gap', *Centre for International Gap Innovation* (2022), *www.cigionline.org/articles/ai-and-the-actual-ihl-accountability-gap/* (accessed 15 August 2023).

Chapter 1

1. British Red Cross, *www.redcross.org.uk/about-us/what-we-do/protecting-people-in-armed-conflict/international-humanitarian-law#Fundamental* (accessed 15 August 2023).

2. Robert H. Jackson Center, 'The Influence of the Nuremberg Trial on International Criminal Law', *https://www.roberthjackson.org/speech-and-writing/the-influence-of-the-nuremberg-trial-on-international-criminal-law/ (accessed 12 June 2024)*.
3. Mark R. Amstutz, *International Ethics: Concepts, Theories, and Cases in Global Politics*, 4th edn (Lanham MD: Rowman and Littlefield Publishers, 2013), p. 8.
4. George F. Kennan, 'Morality and Foreign Policy', *Foreign Affairs*, 64/2 (1985–6), 205–18 (206).
5. A. J. Coates, *The Ethics of War* (Manchester: Manchester University Press, 2016), p. 35.
6. Coates, *The Ethics of War*, p. 36.
7. Hans Morgenthau, *Politics Among Nations: The Struggle for Power and Peace* (New York: Alfred A. Knopf, 1973), p. 13.
8. Morgenthau, *Politics Among Nations*, p. 11.
9. Ernest W. Lefever, *Ethics and World Politics* (Washington DC: Ethics and Public Policy Center, 1988), p. 27.
10. E. H. Carr, *The Twenty Years Crisis, 1919–1939* (London: Macmillan, 1981), p. 153.
11. Carl Von Clausewitz, *On War* (London: Penguin, 1982), p. 103.
12. Clausewitz, *On War*, p. 102.
13. Brian Orend, *War and Political Theory* (Cambridge: Polity, 2019), p. 47.
14. Coates, *The Ethics of War*, p. 46.
15. Bernard Law Montgomery, *Memoirs* (London: Collins, 1958), pp. 88–9.
16. Coates, *The Ethics of War*, p. 48.
17. Robin W. Lovin, *Reinhold Niebuhr and Christian Realism* (Cambridge: Cambridge University Press, 1995).
18. Hans Morgenthau, *Scientific Man vs. Power Politics* (London: Latimer House, 1947), p. 203.
19. Joseph Evans and Leo R. Ward (eds), *The Social and Political Philosophy of Jacques Maritain* (London: Geoffrey Bles, 1956), p. 320.
20. Marshall Cohen, Thomas Nagel and Thomas Scanlon (eds), *War and Moral Responsibility* (Princeton NJ: Princeton University Press, 1974), p. 70.
21. Orend, *War and Political Theory*, p. 50.
22. John Mearsheimer, *The Tragedy of Great Power Politics* (New York, London: Norton, 2001), p. 3.
23. Morgenthau, *Scientific Man vs. Power Politics*.
24. Hans Morgenthau, *Decline of Domestic Politics* (Chicago IL: University of Chicago Press, 1958).
25. Richard N. Lebow, 'Classical Realism', in T. Dunne, M. Kurki and S. Smith (eds), *International Relations Theories, Discipline and Diversity*, 4th edn (Oxford: Oxford University Press, 2016), p. 42.

26 Raymond Aron, *Peace & War: A Theory of International Relations* (New Brunswick: Transaction Publishers, 2009), p. 588.
27 Donald MacKenzie, 'Militarism and Socialist Theory', *Capital and Class*, 7/1 (1983), 35.
28 Jean Dreze, 'Militarism, Development and Democracy', *Economic and Political Weekly*, 35/14 (2000), 1171.
29 Ulrich Albrecht, Dieter Ernst, Peter Lock and Herbert Wulf, 'Militarization, Arms Transfer and Arms Production in Peripheral Countries', *Journal of Peace Research*, 3/12 (1975), 195–212; Cynthia Enloe, *Does Khaki Become You? The Militarization of Women's Lives* (London: Pluto Press, 1988); Edward Palmer Thompson, 'Notes on Extremism: The Last Stage of Civilization', in *Exterminism and Cold War*, New Left Review (London: Verso, 1982), pp. 1–33.
30 Thompson, 'Notes on Extremism'.
31 Michael W. Doyle, 'Kant, Liberal Legacies and Foreign Affairs', *Philosophy and Public Affairs*, 12/4 (1983), 323–53.
32 William Blum, *Rogue State: A Guide to the World's Only Superpower*, 3rd edn (London: Zed, 2006).
33 SIPRI, 'World military spending reached $1.6 trillion in 2010, biggest increase in South America, fall in Europe according to new SIPRI data' (11 April 2011), *www.sipri.org/media/pressrelease/milex* (accessed 15 August 2023).
34 US Department of Defense, 'Active Duty Military Strength and Other Personnel Statistics (2011), *https://download.militaryonesource. mil/12038/MOS/Reports/2011_Demographics_Report.pdf* (accessed 15 August 2023).
35 Michael Mann, 'Authoritarian and Liberal Militarism: A Contribution from Comparative and Historical Sociology', in S. Smith, K. Booth and M. Zalewski (eds), *International Theory: Positivism and Beyond* (Cambridge: Cambridge University Press, 1996), pp. 221–39.
36 Alfred Vagts, *A History of Militarism: Civilian and Military*, 2nd edn (New York: Meridian Books, 1959).
37 Volker R. Berghahn, *Militarism: The History of an International Debate 1861–1979* (Cambridge: Cambridge University Press, 1981), p. 32.
38 Michael Mann, 'The Roots and Contradictions of Modern Militarism', *New Left Review*, 162 (1987), *https://newleftreview.org/issues/ i162/articles/michael-mann-the-roots-and-contradictions-of-modern-militarism.pdf* (accessed 12 June 2024).
39 Enloe, *Does Khaki Become You?*
40 A. Stavrianakis and J. Selby (eds), *Militarism and International Relations: Political Economy, Security, Theory* (London: Routledge, 2013).

41 Asbjørn Eide and Marek Thee (eds), *Problems of Contemporary Militarism* (London: Croom Helm, 1980) p. 9.
42 David Kinsella, 'The Global Arms Trade and the Diffusion of Militarism', in Stavrianakis and Selby (eds), *Militarism and International Relations*, p. 116.
43 Andrew L. Ross, 'Dimensions of Militarization in the Third World', *Armed Forces and Society*, 13/4 (1987), 561–78.
44 Martin Shaw, *War and Genocide: Organized Killing in Modern Society* (Cambridge: Polity, 2003), p. 106.
45 James Eastwood, *Ethics as a Weapon of War: Militarism and Morality in Israel* (Cambridge: Cambridge University Press, 2020), p. 3.
46 Martin Shaw, *Post-Military Society: Militarism, Demilitarisation, and War at the End of the Twentieth Century* (Philadelphia PA: Temple University Press, 1991); Martin Shaw, 'Twenty-First Century Militarism: A Historical-Sociological Framework', in Stavrianakis and Selby (eds), *Militarism and International Relations*, pp. 19–32.
47 Uri Ben-Eliezer, *The Making of Israeli Militarism* (Bloomington IN: Indiana University Press, 1998).
48 Louis Althusser, 'Ideology and Ideological State Apparatuses', in *Lenin and Philosophy and Other Essays* (New York: Monthly Review Press, 1971), https://www.csun.edu/~snk1966/Lous%20Althusser%20 Ideology%20and%20Ideological%20State%20Apparatuses.pdf (accessed 12 June 2024).
49 Eastwood, *Ethics as a Weapon of War*.
50 Martin Ceadel, *Thinking about Peace and War* (Oxford: Oxford University Press, 1987).
51 Alfred Vagts, *A History of Militarism: Romance and Reality of a Profession* (New York: W. W. Norton and Co., 1973), p. 11.
52 Coates, *The Ethics of War*, p. 60.
53 Friedrich Von Bernhardi, *Deutschland und der Nächste Krieg* ('*Germany and the Next War*') trans. by Allen H. Powless (London: Edward Arnold, 1912), p. 31.
54 Max Scheler, *Der Genius Des Krieges Und Der Deutsche Krieg* (London: Wentworth Press 1918), p. 46.
55 Anthony M. Ludovici, *A Defence of Conservatism: A Further Textbook for Tories* (London: Faber & Gwyer, 1927), p. 253, www.anthony mludovici.com/dc_01.htm (accessed 15 August 2023).
56 Ceadel, *Thinking about Peace and War*, p. 8.
57 Ceadel, *Thinking about Peace and War*, p. 45.
58 Coates, *The Ethics of War*, p. 64.
59 Paul Fussell, *Wartime: Understanding and Behaviour in the Second World War* (New York: Oxford University Press, 1989), pp. 166–7.
60 Aron, *Peace & War*, p. 593.

61 Coates, *The Ethics of War*, p. 82.
62 Julien Benda, *The Treason of the Intellectuals* (New York: W. W. Norton & Company, 1969).
63 Shaw, *Post-Military Society*, p. 29.
64 Shaw, *Post-Military Society*, p. 30.
65 Derek Gregory, 'War and Peace', *Transactions of the Institute of British Geographers*, 35/2 (2010), 168.
66 Anna Stavrianakis and Jan Selby, 'War Becomes Academic: Human Terrain, Virtuous War and Contemporary Militarism. An Interview with James Der Derian', in Stavrianakis and Selby (eds), *Militarism and International Relations*, p. 68.
67 Coates, *The Ethics of War*, p. 115.
68 Coates, *The Ethics of War*, p. 116.
69 Michael Walzer, *Just and Unjust Wars*, 3rd edn (New York: Basic Books, 2000), p. 21.
70 Anthony F. Lang Jr, Cian O'Driscoll and John Williams (eds), *Just War: Authority, Tradition, and Practice* (Washington DC: Georgetown University Press, 2013), p. 1.
71 Orend, *War and Political Theory*, p. 80.
72 Orend, *War and Political Theory*, pp. 86–87.
73 International Committee of the Red Cross, 'What are *jus ad bellum* and *jus in bello*?' (2015), https://www.icrc.org/en/document/what-are-jus-ad-bellum-and-jus-bello-0%EF%BB%BF (accessed 12 June 2024).

Chapter 2

1 David Jordan, 'Air and Space Warfare', in D. Jordan, J. D. Kiras, D. J. Lonsdale, I. Speller, C. Tuck and C. D. Walton (eds), *Understanding Modern Warfare* (Cambridge: Cambridge University Press, 2008), p. 180.
2 Lily Hamourtziadou and Jonathan Jackson, 'Winning Wars: The Myths and Triumphs of Technology', *The Journal of Global Faultlines*, 6/2 (2019), 127–38.
3 Jordan, 'Air and Space Warfare', p. 198.
4 National Museums Liverpool, https://www.liverpoolmuseums.org.uk/museum-of-liverpool/blitz/stories-liverpool-blitz#section--margaret-johnson (accessed 12 June 2024).
5 A. C. Grayling, *Among the Dead Cities* (London: Bloomsbury, 2006), p. 18.
6 W. G. Sebald, *On the Natural History of Destruction* (London: Random House, 2003), p. 29.
7 Martin Middlebrook, *The Battle of Hamburg: The Firestorm Raid* (London: Allen Lane, 1980), p. 276.

8 World War II Database, *https://ww2db.com/battle_spec.php?battle_id=55#:~:text=RAF%20bombers%20attacked%20targets%20in,Hanover%2C%20and%20Emden%2C%20Germany.&text=The%20British%20RAF%20Bomber%20Command,%2C%20Germany%2C%20causing%20minimal%20damage.&text=British%20bombers%20attacked%20Hamburg%20and%20Berlin%20in%20Germany%2C%20causing%20heavy%20casualties* (accessed 15 August 2023).
9 Igor Primoratz (ed.), *Terror from the Sky: The Bombing of German Cities in World War II* (Oxford: Berghahn Books, 2014).
10 ICAN, 'Hiroshima and Nagasaki bombings', *https://www.icanw.org/hiroshima_and_nagasaki_bombings* (accessed 12 June 2024).
11 International Humanitarian Law Databases, 'Final Act of the International Peace Conference. The Hague, 29 July 1899', International Committee of the Red Cross, *https://ihl-databases.icrc.org/en/ihl-treaties/hague-finact-1899* (accessed 15 August 2023).
12 Grayling, *Among the Dead Cities*, p. 146.
13 Stanley Baldwin, 'A Fear for the Future' (1932), *https://en.wikisource.org/wiki/A_Fear_For_The_Future* (accessed 15 August 2023).
14 Baldwin, 'A Fear for the Future'.
15 UK Parliament, 'Bombing of Civil Populations', *https://hansard.parliament.uk/commons/1940-02-06/debates/b33adb9d-dbcf-485c-b102-44abd743ab14/BombingOfCivilPopulations* (accessed 15 August 2023).
16 Grayling, *Among the Dead Cities*, 254.
17 Robin Neillands, *The Bomber War: The Allied Air Offensive Against Germany* (New York: Overlook Press, 2001), p. 386.
18 Clausewitz and Macaulay cited in Neillands, *The Bomber War*, p. 386.
19 Grayling, *Among the Dead Cities*, p. 212.
20 Grayling, *Among the Dead Cities*, p. 215.
21 Matthew M. Caverly, 'Nazi Militarism', *Library of Social Science* (2016), *www.libraryofsocialscience.com/newsletter/posts/2016/2016-12-22-caverly.html* (accessed 15 August 2023).
22 Hajo Holborn, *The History of Modern Germany, 1840–1945* (Princeton NJ: Princeton University Press, 1969).
23 Caverly, 'Nazi Militarism'.
24 R. Koenigsberg, *Nations have the Right to Kill: Hitler, the Holocaust, and War* (New York: Library of Social Science, 2009).
25 Matthew Cooper, *The German Army, 1933–45* (Chelsea MI: Scarborough House, 1990); Congressional Budget Office, 'U.S. Defense Spending, Post-Cold War', *Government Report* (Washington, DC: GPO, 2010.)

26 Heinrich Von Treitschke, *Treitschke's History of Germany in the 19th Century*, 7 vols, ed. and trans. by P. Eden and P. Cedar (London: Jarrold and Sons, G. Allen and Unwin, 1915–19).
27 Laurence Rees, *Auschwitz: The Nazis and the 'Final Solution'* (London: Penguin Random House, 2005), p. 171.
28 Rees, *Auschwitz*, p. 172.
29 Holocaust Memorial Day Trust, *www.hmd.org.uk/resource/22-march-1943-large-gas-chambers-at-auschwitz-birkenau-come-into-operation/* (accessed 15 August 2023).
30 Rees, *Auschwitz*, p. 198.
31 Holocaust Encyclopedia, 'Sonderkommandos', *https://encyclopedia.ushmm.org/content/en/article/sonderkommandos* (accessed 15 August 2023).
32 Rees, *Auschwitz*, p. 290.
33 Jennifer Rosenberg, 'Zyklon B, a Poison Used During the Holocaust' (23 January 2020), *www.thoughtco.com/zyklon-b-gas-chamber-poison-1779688*; A. L. Ross, 'Dimensions of Militarization in the Third World', *Armed Forces and Society*, 13/4 (1987), 561–78.
34 Mary L. Cummings, 'Ethical and Social Issues in the Design of Weapon Control Computer Interfaces', *Proceedings: International Symposium on Technology and Society, 2003* (Crime Prevention, Security and Design) (2003) 14–18, *https://ieeexplore.ieee.org/xpl/conhome/9312/proceeding* (accessed 15 August 2023).
35 Abraham M. Rutchick, Ryan M. McManus, Denise M. Barth, Robert J. Youmans, Andrew T. Ainsworth and Johnny H. Goukassian, 'Technologically Facilitated Remoteness Increases Killing Behavior', *Journal of Experimental Social Psychology*, 73 (2017), 147–50. Saint Augustine, 'City of God', *Logos Virtual Library*, *www.logoslibrary.org/augustine/city/0105.html* (accessed 15 August 2023).
36 Carmen Chandler, 'It is Psychologically Easier to Kill from a Distance, According to New CSUN Study', *CSUN Today* (2017), *https://csunshinetoday.csun.edu/media-releases/it-is-psychologically-easier-to-kill-from-a-distance-according-to-new-csun-study/* (accessed 15 August 2023).
37 Chandler, 'It is Psychologically Easier to Kill from a Distance'.
38 Chandler, 'It is Psychologically Easier to Kill from a Distance'.
39 Tom Jacobs, 'Physical Remoteness Makes Killing Easier', *Pacific Standard* (13 July 2017), *https://psmag.com/news/physical-remoteness-makes-killing-easier* (accessed 15 August 2023).
40 Jitender Sareen, Brian J. Cox, Tracie O. Afifi, Murray B. Stein, Shay-Lee Belik, Asmundson Meadows and J. G. Gordon, 'Combat and Peacekeeping Operations in Relation to Prevalence of Mental Disorders and

Perceived Need for Mental Health Care', *Arch Gen Psychiatry*, 64/7 (2007), 843–52.

41 Shira Maguen, David D. Luxton, Nancy A. Skopp, Gregory A. Gahm, Mark A. Reger, Thomas J. Metzler, Charles R. Marmar, 'Killing in Combat, Mental Health Symptoms, and Suicidal Ideation in Iraq War Veterans', *Journal of Anxiety Disorders*, 25/4 (2011), 563–67.

42 Dave Grossman, *On Killing: The Psychological Cost of Learning to Kill in War and Society* (New York: Back Bay Books, 2009).

43 Stanley Milgram, *Obedience to Authority: An Experimental View* (New York: Harper and Row, 1974).

44 Joshua D. Greene, Brian R. Sommerville, Leigh E. Nystrom, John M. Darley and Jonathan D. Cohen, 'An fMRI investigation of emotional engagement in moral judgment', *Science*, 293/5537 (2001), 2105–8.

45 Errol A. Henderson, 'Hidden in Plain Sight: Racism in International Relations Theory', in A. Anievas, N. Manchanda and R. Shilliam (eds) *Race and Racism in International Relations: Confronting the Global Colour Line* (Abingdon: Routledge, 2014), p. 20.

46 Cynthia Enloe, *Globalization and Militarism: Feminists Make the Link* (Lanham MD: Rowman & Littlefield, 2007), p. 11.

47 James Eastwood, 'Rethinking Militarism as Ideology: The Critique of Violence After Security', *Security Dialogue*, 49/1–2 (2018), 48.

48 Jasmine K. Gani, 'Racial Militarism and Civilizational Anxiety at the Imperial Encounter: From Metropole to the Postcolonial State', *Security Dialogue*, 52/6 (2021), 546.

49 Gani, 'Racial Militarism and Civilizational Anxiety at the Imperial Encounter', 546.

50 Bernard Williams, *Ethics and the Limits of Philosophy* (London: Routledge, 2006), p. 204.

51 Hans-Jörg Sigwart, 'The Logic of Legitimacy: Ethics in Political Realism', *The Review of Politics*, 75/3 (2013), 407–32.

52 John Gray, *Black Mass: Apocalyptic Religion and the Death of Utopia* (London: Penguin, 2008), 272–3.

53 John Gray, *Heresies: Against Progress and Other Illusions* (London: Granta Books, 2004), p. 3.

Chapter 3

1 Peter Lee, *Reaper Force* (London: John Blake Publishing, 2018), p. 1.
2 Neillands, *The Bomber War*, p. 12.
3 Curtis LeMay and McKinley Kantor, *Mission with LeMay* (New York: Garden City Press, 1965), p. 383.
4 Neillands, *The Bomber War*, p. 95.

5. David Hambling, *Swarm Troopers: How Small Drones will Conquer the World* (Venice FL, Archangel Ink, 2015), p. 11.
6. Brett Holman, 'The First Air Bomb: Venice, 15 July 1849' (2009), https://airminded.org/2009/08/22/the-first-air-bomb-venice-15-july-1849/ (accessed 15 August 2023).
7. Norman Polmar, 'An Early Pilotless Aircraft', *Naval History Magazine*, 33/4 (2019), https://www.usni.org/magazines/naval-history-magazine/2019/august/early-pilotless-aircraft (accessed 12 June 2024).
8. Hambling, *Swarm Troopers*, pp. 27–8.
9. Allen T. Duffin, 'Project Fugo: The Japanese Balloon Bombs', https://warfarehistorynetwork.com/project-fugo-the-japanese-balloon-bombs/ (accessed 15 August 2023).
10. 'The dronefather', *The Economist* (1 December 2012), www.economist.com/technology-quarterly/2012/12/01/the-dronefather (accessed 15 August 2023).
11. 'History of Drone Warfare', *The Bureau of Investigative Journalism*, www.thebureauinvestigates.com/explainers/history-of-drone-warfare (accessed 15 August 2023).
12. Charles Perkins, 'U.S.-Israel Strategic Cooperation: Israeli Drones Support U.S. Operations in Kosovo', *Jewish Virtual Library* (1999), www.jewishvirtuallibrary.org/israeli-drones-support-u-s-operations-in-kosovo (accessed 15 August 2023).
13. Filip Rudic, 'Serbia Mourns 1999 Bombing Victims; Kosovo Thanks NATO', *Balkan Transitional Justice* (24 March 2017), https://balkaninsight.com/2017/03/24/serbia-marks-18th-anniversary-of-nato-bombing-03-24-2017/ (accessed 15 August 2023).
14. UN International Criminal Tribunal for the former Yugoslavia, 'Final Report to the Prosecutor by the Committee Established to Review the NATO Bombing Campaign Against the Federal Republic of Yugoslavia' (6 June 2000), http://www.icty.org/x/file/Press/nato061300.pdf (accessed 15 August 2023).
15. UN International Criminal Tribunal for the former Yugoslavia, 'Final Report to the Prosecutor by the Committee Established to Review the NATO Bombing Campaign Against the Federal Republic of Yugoslavia', p. 2.
16. Bill Yenne, *Attack of the Drones: A History of Unmanned Aerial Combat* (London: Zenith Press, 2004), p. 85.
17. White House Press release, 'President Speaks at FBI on New Terrorist Threat Integration Center' (14 February 2003), https://irp.fas.org/news/2003/02/wh021403b.html (accessed 15 August 2023).
18. Bob Woodward, *Obama's Wars* (New York: Simon and Schuster, 2010), p. 6.
19. Peter W. Singer, *Wired for War* (New York: Penguin Books, 2010), p. 4.

20 Lawrence Tritle, 'Warfare in Herodotus', in C. Dewald and J. Marincola (eds), *The Cambridge Companion to Herodotus* (Cambridge: Cambridge University Press, 2006), p. 209.
21 Singer, *Wired for War*, p. 6.
22 Forbes, '30 Great Moments in the History of Robots' (2012), *www.forbes.com/sites/davidewalt/2012/11/27/30-great-moments-in-the-history-of-robots/?sh=6e4c31c1534b* (accessed 15 August 2023).
23 Singer, *Wired for War*, p. 10.
24 Singer, *Wired for War*, pp. 56–7.
25 Singer, *Wired for War*, p. 63.
26 Singer, *Wired for War*, p. 64.
27 John W. Clark, 'Remote Control in Hostile Environments', *New Scientist*, 22/389 (1964), 300.
28 Singer, *Wired for War*, p. 64.
29 Daniel Nelson, 'Mach Speed: From Mach 1 To Mach 3 Speed and Beyond', *Science Trends* (2017), *https://sciencetrends.com/mach-speed-breakdown-examples-mach-1-2-3-beyond/* (accessed 15 August 2023).
30 Gregoire Chamayou, *Drone Theory* (New York: Penguin Books, 2015), pp. 21–2.
31 Chamayou, *Drone Theory*, p. 23.
32 Kenneth Anderson, 'Rise of the Drones: Unmanned Systems and the Future of War', *Written Testimony Submitted to Subcommittee on National Security and Foreign Affairs, Committee on Oversight and Government Reform, US House of Representatives, Subcommittee Hearing* (23 March 2010), p. 12.
33 Bradley J. Strawser, 'The Morality of Drone Warfare Revisited', *The Guardian* (6 August 2012).
34 Scott Shane, S., 'The Moral Case for Drones', *New York Times* (14 July 2012).
35 Chamayou, *Drone Theory*, p. 141.
36 Singer, *Wired for War*, p. 396.
37 Jane Mayer, 'The Predator War', *New Yorker* (26 October 2009).
38 Stanley Milgram, *Obedience to Authority: An Experimental View* (New York: Harper and Row, 1974), p. 39.
39 Milgram, *Obedience to Authority*, p. 38.
40 Chamayou, *Drone Theory*, p. 119.
41 Hambling, *Swarm Troopers*, p. 51.
42 Singer, *Wired for War*, p. 76.
43 Chamayou, *Drone Theory*, p. 17.
44 Singer, *Wired for War*, p. 323.
45 Chamayou, *Drone Theory*, p. 143.
46 Joint Doctrine Note 2/11, 'The UK Approach to Unmanned Aircraft Systems, Ministry of Defence' (2011), p. 56, *https://assets.publishing.*

service.gov.uk/government/uploads/system/uploads/attachment_data/file/644084/20110505-JDN_2-11_UAS_archived-U.pdf (accessed 15 August 2023).
47. Asa Kasher and Amos Yadlin, 'Military Ethics of fighting Terror: An Israeli Perspective', *Journal of Military Ethics*, 4/1 (2005), 20.
48. Hambling, *Swarm Troopers*, pp. 51–2.
49. Immanuel Kant, *Political Writings* (Cambridge: Cambridge University Press, 2007), p. 168.
50. Hambling, *Swarm Troopers*, p. 52.
51. WikiLeaks, 'CIA Best Practices in Counterinsurgency' (18 December 2014), https://wikileaks.org/cia-hvt-counterinsurgency/WikiLeaks_Secret_CIA_review_of_HVT_Operations.pdf (accessed 15 August 2023).
52. William James, 'The Moral Equivalent of War', in L. Bramson and G. W. Goethals (eds), *War* (New York: Basic Books, 1964), p. 24.
53. James, 'The Moral Equivalent of War', p. 27.
54. James, 'The Moral Equivalent of War', p. 28.
55. Anthony W. Marx, *Faith in Nation: Exclusionary Origins of Nationalism* (New York: Oxford University Press, 2005), p. 6.
56. Tyler Wall and Torin Monahan, 'Surveillance and violence for afar: the politics of drones and liminal security-scapes', *Theoretical Criminology*, 15/3 (2011), 246.
57. Kevin Robins and Les Levidow, 'Socializing the Cyborg Self: The Gulf War and Beyond', in C. H. Gray (ed.), *The Cyborg Handbook* (New York: Routledge, 1995), p. 121.
58. E. F. M. Durbin and John Bowlby, 'Personal Aggressiveness and War', in L. Bramson and G. W. Goethals (eds), *War* (New York, London: Basic Books, 1964), p. 81.
59. Durbin and Bowlby, 'Personal Aggressiveness and War', p. 82.
60. Charlie Savage, Eric Schmitt, Azmat Khan, Evan Hill and Christoph Koettl, 'Newly Declassified Video Shows U.S. Killing of 10 Civilians in Drone Strike', *New York Times* (19 January 2022), www.nytimes.com/2022/01/19/us/politics/afghanistan-drone-strike-video.html (accessed 15 August 2023).
61. Larry May, *War Crimes and Just War* (Cambridge: Cambridge University Press, 2007), p. 268.
62. May, *War Crimes and Just War*, p. 206.
63. May, *War Crimes and Just War*, p. 207.
64. BBC, 'Qasem Soleimani: US Strike on Iran General was Unlawful, UN Expert Says' (9 July 2020), www.bbc.co.uk/news/world-middle-east-53345885 (accessed 15 August 2023).
65. UN, 'All drone strikes "in self-defence" should go before Security Council, argues independent rights expert' (9 July 2020), https://news.un.org/en/story/2020/07/1068041 (accessed 15 August 2023).

66 Mia Swart, 'Death by Drone: How can states justify targeted killings?', *Al Jazeera* (11 July 2020), *www.aljazeera.com/news/2020/7/11/death-by-drone-how-can-states-justify-targeted-killings* (accessed 15 August 2023).
67 Simone Weil, *Formative Writings* (New York: Routledge, 1999), p. 173.
68 Christina Balis and Paul O'Neill, *Trust in AI: Rethinking Future Command* (Royal United Services Institute for Defence and Security Studies, Occasional Paper, 2022), p. 5, *www.qinetiq.com/-/media/59de8c79f10d4c5699b5a1f25f2d2279.ashx* (accessed 15 August 2023).
69 Mariarosaria Taddeo, 'Trusting Digital Technologies Correctly', *Minds and Machines*, 27/4 (2017), 565–8.
70 Balis and O'Neill, *Trust in AI: Rethinking Future Command*, pp. 6–7.
71 David Kilcullen, *The Dragons and the Snakes: How the Rest Learned to Fight the West* (London: Hurst, 2022), p. 119.
72 Jean-Baptiste Jeangene Vilmer, 'France and the American Drone Precedent', in Daniel R. Brunstetter R. and J.-V. Holeindre (eds), *The Ethics of War and Peace Revisited: Moral Challenges in an Era of Contested and Fragmented Sovereignty* (Washington DC: Georgetown University Press, 2018), p. 99.
73 Vilmer, 'France and the American Drone Precedent', p. 110.

Chapter 4

1 Andrew Cockburn, *Kill Chain: Drones and the Rise of High-Tech Assassins* (London: Verso, 2016), p. 145.
2 Laurie Calhoun, *We Kill Because We Can* (London: Zed Books, 2015), p. 13.
3 Calhoun, *We Kill Because We Can*, p. 43.
4 Lily Hamourtziadou, *Body Count: The War on Terror and Civilian Deaths in Iraq* (Bristol: Bristol University Press, 2021).
5 Francis Fukuyama, 'The End of History', *National Interest*, 16 (1989), 3–18.
6 Charles Krauthammer, 'Universal Dominion: Toward a Unipolar World', *National Interest*, 18 (1989), 48–9.
7 Charles Krauthammer, 'The War: A Road Map', *Washington Post* (28 September 2001).
8 Charles Krauthammer, 'Coming Ashore: The War in Not Just to Disarm Saddam. It is to Reform a Whole Part of the World', *Time* (17 February 2003).
9 Gary Dorrien, *Imperial Designs: Neoconservatism and the New Pax Americana* (New York: Routledge, 2004), p. 155.

10 Vassilis Fouskas and Bulent Gokay, *The New American Imperialism: Bush's War on Terror and Blood for Oil* (Westport CT: Praeger Security International, 2005), p. 126.
11 Pater Scott, *The Road to 9/11: Wealth, Empire and the Future of America* (Los Angeles CA: University of California Press, 2007), p. 187.
12 Robert Kagan and William Kristol, 'Right War', *Weekly Standard* (1 October 2001), *https://carnegieendowment.org/2001/10/01/right-war-pub-791* (accessed 15 August 2023).
13 George W. Bush, 'Address to Citadel Cadets' (11 December 2001), *www.americanrhetoric.com/speeches/gwbushcitadelcadets.htm* (accessed 15 August 2023).
14 Project on Defense Alternatives, 'Operation Enduring Freedom: Why a Higher Rate of Civilian Bombing Casualties', *Briefing Report #13* (revised 24 January 2002), *www.comw.org/pda/0201oef.html* (accessed 15 August 2023).
15 Project on Defense Alternatives, 'Strange Victory: A Critical Appraisal of Operation Enduring Freedom and the Afghanistan War', *Research Monograph #6* (30 January 2002), *www.comw.org/pda/0201strangevic.html* (accessed 15 August 2023).
16 Lily Hamourtziadou, 'Iraq 20 Years On: death came from the skies on March 19 2003 – and the killing continues to this day', *The Conversation* (17 March 2023), *https://theconversation.com/iraq-20-years-on-death-came-from-the-skies-on-march-19-2003-and-the-killing-continues-to-this-day-201988* (accessed 15 August 2023).
17 Iraq Body Count database, *www.iraqbodycount.org/database/incidents/x001* (accessed 5 February 2024).
18 Iraq Body Count database, *www.iraqbodycount.org/database/incidents/x002* (accessed 5 February 2024).
19 Iraq Body Count database, *www.iraqbodycount.org/database/incidents/x003* (accessed 5 February 2024).
20 Iraq Body Count database, *www.iraqbodycount.org/database/incidents/x038* (accessed 5 February 2024).
21 Iraq Body Count database, *www.iraqbodycount.org/database/incidents/a6384* (accessed 5 February 2024).
22 Cockburn, *Kill Chain*, pp. 137–8.
23 Cockburn, *Kill Chain*, p. 138.
24 Cockburn, *Kill Chain*, p. 139.
25 Iraq Body Count, 'A Dossier of Civilian Casualties 2003–2005' (2005), *www.iraqbodycount.org/analysis/reference/pdf/a_dossier_of_civilian_casualties_2003-2005.pdf* (accessed 15 August 2023).
26 Iraq Body Count, *www.iraqbodycount.org/database/* (accessed 5 February 2024).

27 Muhammad Idrees Ahmad, 'Death from Above, Remotely Controlled: Obama's Drone Wars', *In These Times* (2015), https://inthesetimes.com/article/drones-andrew-cockburn-kill-chain-chris-woods-sudden-justice (accessed 12 June 2024).
28 David Rohde, 'The Drone Wars', *Reuters* (26 January 2012), https://www.reuters.com/article/idUSTRE80P11M/ (accessed 12 June 2024).
29 Cockburn, *Kill Chain*, p. 227.
30 Iona Craig, 'What Really Happened when a US Drone Hit a Yemeni Wedding Convoy?', *Aljazeera* (20 January 2014), http://america.aljazeera.com/watch/shows/america-tonight/america-tonight-blog/2014/1/17/what-really-happenedwhenausdronehitayemeniweddingconvoy.html (accessed 15 August 2023).
31 Iraq Body Count, *www.iraqbodycount.org/database/* (accessed 5 February 2024).
32 Iraq Body Count database, *www.iraqbodycount.org/database/incidents/a5888* (accessed 5 February 2024).
33 Iraq Body Count database, *www.iraqbodycount.org/database/incidents/a5885* (accessed 5 February 2024).
34 Airwars, 'Tens of Thousands of Civilians Likely Killed by US in "Forever Wars"' (2021), https://airwars.org/investigations/tens-of-thousands-of-civilians-likely-killed-by-us-in-forever-wars/ (accessed 15 August 2023).
35 Iraq Body Count database, *www.iraqbodycount.org/database/incidents/a5878a* (accessed 5 February 2024).
36 Iraq Body Count database, *www.iraqbodycount.org/database/incidents/a6317* (accessed 5 February 2024).
37 Krishnadev Calamur, 'Casualties in the Fight Against ISIS', *The American* (4 May 2016), www.theatlantic.com/national/archive/2016/05/american-death-toll-isis/481206/ (accessed 15 August 2023).
38 James Kiras, *Special Operations and Strategy: From World War II to the War on Terrorism* (Abingdon: Routledge, 2006), p. 12.
39 Thomas Waldman, *Vicarious Warfare: American Strategy and the Illusion of War on the Cheap* (Bristol: Bristol University Press, 2021), p. 119.
40 Waldman, *Vicarious Warfare*, p. 122.
41 J. Marshall Beier, 'Outsmarting Technologies; Rhetoric, Revolutions in Military Affairs, and the Social Depth of Warfare', *International Politics*, 43/2 (2006), 266–80.
42 Victor Davis Hanson, *Father of Us All: War and History, Ancient and Modern* (New York: Bloomsbury, 2010), p. 11.
43 James Holmes, 'Warmaking by Remote Control is a False Choice', *The National Interest* (25 November 2019), https://nationalinterest.org/blog/warmaking-remote-control-false-choice-99007 (accessed 15 August 2023).

44 Waldman, *Vicarious Warfare*, p. 130.
45 Amnesty International, 'Somalia: Zero Accountability as Civilian Deaths Mount from US Air Strikes' (2020), *www.amnesty.org/en/latest/news/2020/04/somalia-zero-accountability-as-civilian-deaths-mount-from-us-air-strikes/* (accessed 15 August 2023).
46 Amnesty International, 'Somalia: Zero accountability as civilian deaths mount from US air strikes' 2020.
47 Waldman, *Vicarious Warfare*, p. 187.
48 Douglas Porch, 'Expendable Soldiers', *Small Wars & Insurgencies*, 25/3 (2014), 696–716.
49 Lydia Day, Frank Ledwidge, Stuart Casey-Maslen and Mark Goodwin-Hudson, 'Avoiding Civilian Harm in Partnered Military Operations: The UK's Responsibility', *Ceasefire Centre for Civilian Rights* (2023), 4.
50 Waldman, *Vicarious Warfare*, p. 180.
51 Waldman, *Vicarious Warfare*, p. 183.
52 John Prados, 'The Continuing Quandary of Covert Operations', *Journal of National Security Law and Policy*, 5/2 (2012), 359–72.
53 Russell A. Burgos, 'Pushing the Easy Button: Special Operations Forces, International Security, and the Use of Force', *Special Operations Journal*, 4/2 (2018), 109–28.
54 Josh Fruhlinger, 'Stuxnet Explained: The First Known Cyberweapon' (2022), *www.csoonline.com/article/3218104/stuxnet-explained-the-first-known-cyberweapon.html#:~:text=Stuxnet%20is%20a%20powerful%20computer,about%20its%20design%20and%20purpose* (accessed 15 August 2023).
55 Lily Hamourtziadou, Hamit Dardagan and John Sloboda, 'Generation: War', *Iraq Body Count* (2018), *www.iraqbodycount.org/analysis/beyond/generation-war/* (accessed 15 August 2023).
56 Andrew J. Bacevich, *The New American Militarism: How Americans are Seduced by War* (Oxford: Oxford University Press, 2013), p. 32.
57 Max Boot, 'American Imperialism? No Need to Run from Label', *USA Today* (6 May 2003).
58 Wright C. Mills, *The Power Elite* (Oxford: Oxford University Press, 1956), p. 184.
59 Bacevich, *The New American Militarism*, p. 147.
60 Barack Obama, 'Obama's Remarks on Military Spending', *New York Times* (5 January 2012), *www.nytimes.com/2012/01/06/us/text-obamas-remarks-on-military-spending.html* (accessed 15 August 2023).
61 Walter Pincus, 'Has Obama taken Bush's Preemption Strategy to Another Level?', *Washington Post* (9 January 2012), *www.washingtonpost.com/world/national-security/has-obama-taken-bushs-preemption-strategy-to-another-level/2012/01/06/gIQAHDImmP_story.html* (accessed 15 August 2023).

62 Bacevich, *The New American Militarism*, p. 232.
63 Calhoun, *We Kill Because We Can*, p. 154.
64 Chalmers Johnson, *The Sorrows of Empire: Militarism, Secrecy, and the End of the Republic* (New York: Metropolitan Books, 2004), p. 4.
65 Johnson, *The Sorrows of Empire*, p. 5.
66 Ronald Steel, *Pax Americana* (New York: Viking, 1967), pp. 17–18.
67 Karen Talbot, 'The Real Reasons for War in Yugoslavia: Backing Up Globalization with Military Might', *Social Justice*, 27/4 (2000), 100.
68 Peter Dizikes, 'Is the *Pax Americana* truly peaceful?', *Massachusetts Institute of Technology* (2017), *https://news.mit.edu/2017/john-dower-book-pax-americana-truly-peaceful-0627* (accessed 15 August 2023).
69 Samuel Moyn, *Humane: How the United States Abandoned Peace and Reinvented War* (London: Verso, 2022), p. 130.
70 Kees Van der Pijl, *The Making of an Atlantic Ruling Class* (New York, London: Verso, 1984), p. 7.
71 William Kristol and Robert Kagan (eds), *Present Dangers: Crisis and Opportunity in American Foreign and Defense Policy* (New York: Encounter Books, 2000).
72 Francis Fukuyama, 'After Neoconservatism', in D. Skidmore (ed.), *Paradoxes of Power: US Foreign Policy in a Changing World* (New York: Routledge, 2007), p. 123.
73 Michael Mastanduno, 'Hegemonic Order, September 11, and the Consequences of the Bush Revolution', *International Relations of the Asia Pacific*, 5 (2005), 177–96.
74 Stuart Kaufman, Richard Little and William C. Wohlforth, *The Balance of Power in World History* (New York: Palgrave Macmillan, 2007), p. 7.
75 John G. Ikenberry and Charles A. Kupchan, 'Socialization and Hegemonic Power', *International Organization*, 44/3 (1990), 283–315.
76 Robert W. Cox, 'Social Forces, States and World Orders: Beyond International Relations Theory', *Millennium: Journal of International Studies*, 10/2 (1981), 126–55.
77 George W. Bush, 'Statement by the President in His Address to the Nation', The White House (11 September 2001), *https://georgewbushwhitehouse.archives.gov/news/releases/2001/09/20010911-16.html* (accessed 15 August 2023).
78 George W. Bush, 'President Bush Announces "Volunteers for Prosperity"', White House (21 May 2003), *http://peacecorpsonline.org/messages/messages/2629/2013673.html* (accessed 15 August 2023).
79 Charles Krauthammer, 'The War: A Road Map', *Washington Post* (28 September 2001).

80 Jeffrey Goldberg, 'The Obama Doctrine', *The Atlantic* (April 2016), p. 233, *www.theatlantic.com/magazine/archive/2016/04/the-obama-doctrine/471525/* (accessed 15 August 2023).
81 Moyn, *Humane*, p. 270.
82 Barack Obama, 'Nobel Lecture' (10 December 2009), *www.nobelprize.org/prizes/peace/2009/obama/lecture/#:~:text=I%20receive%20this%20honor%20with,in%20the%20direction%20of%20justice* (accessed 15 August 2023).
83 John Mearsheimer, *The Tragedy of Great Power Politics* (New York: Norton, 2001), p. 33.
84 Mearsheimer, *The Tragedy of Great Power Politics*, p. 49.
85 Martin Wight, *Power Politics* (London: Royal Institute of International Affairs, 1946), p. 40.
86 Neta C. Crawford, 'Blood and Treasure: United States Budgetary Costs and Human Costs of 20 Years of War in Iraq and Syria, 2003–2023', *Watson Institute International & Public Affairs* (2013), p. 25.
87 Iraq Body Count database, *www.iraqbodycount.org/database/incidents/a6262* (accessed 5 February 2024).
88 Jean-Baptiste Jeangene Vilmer, 'France and the American Drone Precedent', in Daniel R. Brunstetter R. and J.-V. Holeindre (eds), *The Ethics of War and Peace Revisited: Moral Challenges in an Era of Contested and Fragmented Sovereignty* (Washington DC: Georgetown University Press, 2018), p. 99.
89 Vilmer, 'France and the American Drone Precedent', p. 101.
90 Johnson, *The Sorrows of Empire*, p. 285.
91 Robert L. Goldich, 'American Military Culture from Colony to Empire', *Daedalus*, 140/3 (2011), 58–9.
92 Paul Rogers, 'The triple paradigm crisis: economy, environment, and security', *The Journal of Global Faultlines*, 6/2 (2019), 139–49, *www.scienceopen.com/hosted-document?doi=10.13169/jglobfaul.6.2.0139* (accessed 15 August 2023).
93 Antony J. Blinken, 'Remembering the Loss of 13 American Heroes', *US Department of State* (2022), *www.state.gov/remembering-the-loss-of-13-american-heroes/* (accessed 15 August 2023).
94 Amnesty International, 'Killing of Civilians in Basra and al-Amara' (2004), p. 10, *www.amnesty.org/en/wp-content/uploads/2021/06/mde140072004en.pdf* (accessed 15 August 2023).
95 Global Policy Forum, 'Killing of Civilians, Murder and Atrocities' (2007), *https://archive.globalpolicy.org/security/issues/iraq/occupation/report/7atrocities.htm* (accessed 15 August 2023).
96 Gary Younge, 'If Wanton Murder is Essential to the US Campaign in Iraq, it's Time to Leave', *The Guardian* (26 June 2006).

97 Richard Paddock, 'Shots in the Heart of Baghdad', *Los Angeles Times* (25 July 2005).
98 Richard Engel, 'What Happened in Haditha', *NBC News* (30 May 2006).
99 Iraq Body Count database, *www.iraqbodycount.org/database/incidents/k2171* (accessed 5 February 2024).
100 Ellen Knickmeyer and Salih S. Aldin, 'US Raid Kills Family North of Baghdad', *Washington Post* (4 January 2006), *https://www.washingtonpost.com/archive/politics/2006/01/04/us-raid-kills-family-north-of-baghdad-span-classbankheadiraqis-say-12-slain-in-airstrike/a5241600-ff58-4283-9fd7-b9f3cfd1bbfa/* (accessed 12 June 2024).
101 International Criminal Court, 'Situation in Iraq/UK Final Report' (2020), pp. 11–12, *www.icc-cpi.int/sites/default/files/itemsDocuments/201209-otp-final-report-iraq-uk-eng.pdf* (accessed 15 August 2023).
102 Crisis Group Asia, 'Report 326' (2022), *www.crisisgroup.org/asia/south-asia/afghanistan/afghanistans-security-challenges-under-taliban* (accessed 15 August 2023).
103 James Gaffney, 'Just War: The Catholic Contribution to International Law' (paper presented 30 March 2003), *http://cas.loyno.edu/sites/chn.loyno.edu/files/GAFFNEY_JustWar.pdf* (accessed 15 August 2023).
104 Marcus Tullius Cicero, *De Officiis*, eBook, Project Gutenberg (2014), *www.gutenberg.org/files/47001/47001-h/47001-h.htm* (accessed 15 August 2023).
105 Joseph Boyle, 'The Necessity of "Right Intent" for Justifiably Waging War', in Lang Jr, O'Driscoll and Williams (eds), *Just War*, pp. 181–96.
106 Iraq Body Count database, *www.iraqbodycount.org/database/incidents/2014-08-31* (accessed 5 February 2024).
107 United Nations, 'ISIL/Da'esh Committed Genocide of Yazidi, War Crimes against Unarmed Cadets, Military Personnel in Iraq, Investigative Team Head Tells Security Council' (10 May 2021), *https://press.un.org/en/2021/sc14514.doc.htm* (accessed 15 August 2023).
108 Todd A. Burkhardt, 'Just War and Human Rights: Fighting with Right Intention' (PhD thesis, University of Tennessee, Knoxville, 2013), p. 4, *https://trace.tennessee.edu/cgi/viewcontent.cgi?referer=&httpsredir=1&article=2840&context=utk_graddiss#:~:text=A%20state%20pursuing%20a%20just,of%20basic%20justice%20are%20secure* (accessed 15 August 2023).
109 Moyn, *Humane*, p. 134.

Chapter 5

1. Volodymyr Zelenskyy, *A Message from Ukraine* (London: Hutchinson Heinemann, 2022), p. 4.
2. Luke Harding, *Invasion: Russia's Bloody War and Ukraine's Fight for Survival* (London: Guardian Faber, 2022), p. 116.
3. Roberto J. Gonzalez, 'Drones over Ukraine: What the War Means for the Future of Remotely Piloted Aircraft in Combat', *The Conversation* (23 February 2023), *https://theconversation.com/drones-over-ukraine-what-the-war-means-for-the-future-of-remotely-piloted-aircraft-in-combat-197612* (accessed 15 August 2023).
4. Dominika Kunertova, 'The War in Ukraine Shows the Game-Changing Effect of Drones Depends on the Game', *Bulletin of the Atomic Scientists*, 79/2 (2023), 95–102, *https://www.tandfonline.com/doi/full/10.1080/00963402.2023.2178180* (accessed 12 June 2024).
5. Frank De Roose, 'Self-Defence and National Defence', *Journal of Applied Philosophy*, 7/2 (1990), 159–68.
6. Hidemi Suganami, *On the Causes of War* (Oxford: Clarendon Press, 1996), p. 179.
7. Harding, *Invasion*, pp. 4–5.
8. Harding, *Invasion*, p. 5.
9. Harding, *Invasion*, p. 6.
10. Harding, *Invasion*, p. 106.
11. Zelenskyy, *A Message from Ukraine*, pp. 15–16.
12. Zelenskyy, *A Message from Ukraine*, p. 22.
13. Zelenskyy, *A Message from Ukraine*, p. 31.
14. Zelenskyy, *A Message from Ukraine*, p. 45.
15. Zelenskyy, *A Message from Ukraine*, p. 58.
16. Zelenskyy, *A Message from Ukraine*, p. 5.
17. Zelenskyy, *A Message from Ukraine*, p. 6.
18. UN Office of the High Commissioner for Human Rights, 'Ukraine: Civilian Casualty Update 5 June 2023', *https://www.ohchr.org/en/news/2023/06/ukraine-civilian-casualty-update-5-june-2023* (accessed 12 June 2024).
19. Memorial platform tweet (16 May 2023), *https://twitter.com/memorial ua/status/1658522120538791936* (accessed 15 August 2023).
20. Memorial platform tweet (3 June 2023), *https://twitter.com/memorial ua/status/1664874554055880705* (accessed 15 August 2023).
21. Memorial platform tweet (26 May 2023), *https://twitter.com/memorial ua/status/1661975610078511104* (accessed 15 August 2023).
22. Paul Lushenko and Sarah Kreps, 'Americans Support Exporting Drones to Ukraine – with a Caveat', *Brookings* (25 May 2023), *www.brookings.edu/blog/order-from-chaos/2023/05/25/americans-*

support-exporting-drones-to-ukraine-with-a-caveat/ (accessed 15 August 2023).

23. Airforce Technology, 'ALTIUS-600 Small Unmanned Aircraft System' (2021), *www.airforce-technology.com/projects/altius-600-small-unmanned-aircraft-system/* (accessed 15 August 2023).

24. Jack Watling and Nick Reynolds, 'Ukraine at War Paving the Road from Survival to Victory', *Royal United Services Institute* (4 July 2022), *https://static.rusi.org/special-report-202207-ukraine-final-web_0.pdf* (accessed 15 August 2023).

25. Kunertova, 'The War in Ukraine Shows the Game-Changing Effect of Drones Depends on the Game'.

26. Howard Altman, Stetson Payne and Tyler Rogoway, 'Ukraine Unleashes Mass Kamikaze Drone Boat Attack on Russia's Black Sea Fleet Headquarters', *The Drive* (29 October 2022), *www.thedrive.com/the-war-zone/ukraine-unleashes-mass-kamikaze-drone-boat-attack-on-russias-black-sea-fleet-headquarters* (accessed 15 August 2023).

27. RFE/RL, 'Ukraine Says Latest Russian Wave of Drone Strikes Claims 21 Lives in Kherson', *Radio Free Europe, Radio Liberty* (3 May 2023), *www.rferl.org/a/ukraine-russia-bakhmut-drones/32391136.html* (accessed 15 August 2023).

28. Memorial platform tweet (7 June 2023), *https://twitter.com/memorial ua/status/1666324109381320707* (accessed 15 August 2023).

29. Memorial platform tweet (6 June 2023), *https://twitter.com/memorial ua/status/1665962472837206016* (accessed 15 August 2023).

30. A. Moseley, 'Just War Theory', *Internet Encyclopedia of Philosophy*, *https://iep.utm.edu/justwar/* (accessed 15 August 2023).

31. Saint Augustine, 'City of God', *Logos Virtual Library*, *www.logoslibrary.org/augustine/city/0105.html* (accessed 15 August 2023).

32. N. Miščević, 'The Dilemmas of Just War and the Institutional Pacifism', *Revus, Journal for Constitutional Theory and Philosophy of Law*, 13 (2010), *https://journals.openedition.org/revus/1273* (accessed 15 August 2023).

33. Miščević, 'The Dilemmas of Just War and the Institutional Pacifism'.

34. Benjamin Abelow, *How the West Brought War to Ukraine* (Great Barrington MA: Siland Press, 2022), p. 1.

35. Chas Freeman, 'Pushback with Aaron Maté', podcast and video (24 March 2022), *https://thegrayzone.com/2022/03/24/us-fighting-russia-to-the-last-ukrainian-veteran-us-diplomat/* (accessed 15 August 2023).

36. Congressional Research Service, 'U.S. Security Assistance to Ukraine' (2023), *https://crsreports.congress.gov/product/pdf/IF/IF12040?loclr=blogloc* (accessed 15 August 2023).

37. Abelow, *How the West Brought War to Ukraine*, p. 20.

38 John Mearsheimer, 'John Mearsheimer on why the West is Principally Responsible for the Ukrainian Crisis', *The Economist* (19 March 2022), *www.economist.com/by-invitation/2022/03/11/john-mearsheimer-on-why-the-west-is-principally-responsible-for-the-ukrainian-crisis* (accessed 15 August 2023).
39 Brussels Summit Communiqué, press release (14 June 2021), NATO, *www.nato.int/cps/en/natohq/news_185000.htm* (accessed 15 August 2023).
40 Anatoly Antonov, 'An Existential Threat to Europe's Security Architecture?', *Foreign Policy* (2021), *https://foreignpolicy.com/2021/12/30/russia-ukraine-nato-threat-security/* (accessed 15 August 2023).
41 Michael R. Gordon, Bojan Pancevski, Noemie Bisserbe and Marcus Walker, 'Vladimir Putin's 20-Year March to Ukraine – and How the West Mishandled it', *The Wall Street Journal* (1 April 2022), *www.wsj.com/articles/vladimir-putins-20-year-march-to-war-in-ukraineand-how-the-west-mishandled-it-11648826461* (accessed 15 August 2023).
42 Congressional Research Service, 'U.S. Security Assistance to Ukraine' (2023), *https://crsreports.congress.gov/product/pdf/IF/IF12040?loclr=blogloc* (accessed 15 August 2023).
43 Harding, *Invasion*, p. 7.
44 John Mearsheimer, *The Tragedy of Great Power Politics* (New York, London: Norton, 2001), p. 345.
45 Memorial platform tweet (19 June 2023), *https://twitter.com/memorial ua/status/1670822444620304385* (accessed 15 August 2023).
46 Olga Ivshina, Becky Dale and Joseph Lee, 'Counting Russia's Dead in Ukraine – and What it Says about the Changing Face of War', *BBC* (16 June 2023), *www.bbc.co.uk/news/resources/idt-829ea0ba-5b42-499b-ad40-6990f2c4e5d0* (accessed 15 August 2023).
47 Memorial platform tweet (22 June 2023), *https://twitter.com/memorial ua/status/1671806980518731777/photo/3* (accessed 15 August 2023).
48 Rachel Stohl and Suzette Grillot, *The International Arms Trade* (Cambridge: Polity, 2009).
49 Paul Rogers, *Losing Control: Global Security in the Twenty-First Century*, 4th edn (London: Pluto Press, 2021), p. 145.
50 Paul Rogers, 'Our Global Culture of War Means Guaranteed Profits for the Arms Industry', *OpenDemocracy* (23 June 2023), *www.opendemocracy.net/en/arms-industry-shareholder-capitalism-perfect-war-syria-iraq-ukraine/* (accessed 15 August 2023).
51 Katharina Buchholz, 'The World's Biggest Arms Exporters' (2022), *www.statista.com/chart/18417/global-weapons-exports/* (accessed 15 August 2023); R. A. Burgos, 'Pushing the Easy Button: Special Operations Forces, International Security, and the Use of Force', *Special Operations Journal*, 4/2 (2018), 109–28.

52 Zaheena Rasheed, 'How China Became the World's Leading Exporter of Combat Drones', *Aljazeera* (24 January 2003), www.aljazeera.com/news/2023/1/24/how-china-became-the-worlds-leading-exporter-of-combat-drones#:~:text=Data%20from%20the%20Stockholm%20 International,exporter%20of%20the%20weaponised%20aircraft (accessed 15 August 2023).
53 Will Brown, 'Drones Galore: Changing Battlefields', *Centre for Strategic and International Studies* (2022), www.csis.org/analysis/drones-galore-changing-battlefields (accessed 15 August 2023).
54 Andrew Feinstein, *The Shadow World: Inside the Global Arms Trade* (London: Penguin Books, 2012), p. 394.
55 Samir Puri, *Russia's Road to War with Ukraine* (London: Biteback Publishing, 2022), p. 84.
56 Nicholas P. Pacheco, 'How Doctrine and Delineation Can Help Defeat Drones', *Texas National Security Review* (2022), https://waron therocks.com/2022/12/how-doctrine-and-delineation-can-help-defeat-drones/ (accessed 15 August 2023).
57 World of Drones, *New America*, www.newamerica.org/international-security/reports/world-drones/introduction-how-we-became-a-world-of-drones (accessed 15 August 2023).
58 Dionne Searcey, 'Boko Haram is Back. With Better Drones', *New York Times* (23 September 2021), www.nytimes.com/2019/09/13/world/africa/nigeria-boko-haram.html (accessed 15 August 2023).
59 Joanna Frew, *Drone Wars: The Next Generation*, Drone Wars UK (2018).
60 Pacheco, 'How Doctrine and Delineation Can Help Defeat Drones'.
61 Reuters, 'Yemen's Houthis Claim Drone Attack on Refinery in Saudi Capital' (11 March 2022), www.reuters.com/world/middle-east/attack-refinery-riyadh-did-not-affect-petroleum-supplies-spa-2022-03-10/ (accessed 15 August 2023).
62 Frew, *Drone Wars*, pp. 19–20.
63 Paul Adams and Sam Hancock, 'Ben Wallace: Ukraine has "tragically become a battle lab" for War Technology', *BBC* (18 July 2023), www.bbc.co.uk/news/uk-66229336 (accessed 15 August 2023).

Conclusion

1 Amnesty International UK, press release, 'Geneva: No Agreement in Talks on Regulating Killer Robots' (17 December 2021), www.amnesty.org.uk/press-releases/geneva-no-agreement-talks-regulating-killer-robots (accessed 20 August 2023).

NOTES

2. Sam Shead, 'UN Talks to Ban "Slaughterbots" Collapsed – Here's Why That Matters', *CNBC* (22 December 2021), www.cnbc.com/2021/12/22/un-talks-to-ban-slaughterbots-collapsed-heres-why-that-matters.html (accessed 15 August 2023).
3. UN Security Council, 'Final Report of the Panel of Experts on Libya Established Pursuant to Security Council Resolution 1973' (2011), p. 15, https://digitallibrary.un.org/record/3905159?v=pdf (accessed 12 June 2024).
4. UN Security Council, 'Final Report of the Panel of Experts on Libya Established Pursuant to Security Council Resolution 1973', p. 17.
5. Shead, 'UN Talks to Ban "Slaughterbots" Collapsed'.
6. Toby Walsh, 'The Problem with Artificial (General) Intelligence in Warfare', *Centre for Artificial Governance Innovation* (28 November 2022), www.cigionline.org/articles/the-problem-with-artificial-general-intelligence-in-warfare/ (accessed 15 August 2023).
7. Frank Sauer, 'Autonomy in Weapons Systems and the Struggle for Regulation', *Centre for International Governance Innovation* (2022), www.cigionline.org/articles/autonomy-in-weapons-systems-and-the-struggle-for-regulation/ (accessed 15 August 2023).
8. Sauer, 'Autonomy in Weapons Systems and the Struggle for Regulation'.
9. Rebecca Crootof, 'AI and the Actual IHL Accountability Gap', *Centre for International Gap Innovation* (2022), www.cigionline.org/articles/ai-and-the-actual-ihl-accountability-gap/ (accessed 15 August 2023).
10. Eliav Lieblich, 'The Facilitative Function of *Jus in Bello*', *The European Journal of International Law*, 30/1 (2019), 321–40.
11. Human Rights Council, 'Impact of Casualty Recording on the Promotion and Protection of Human Rights', *United Nations General Assembly* (16 May 2023), p. 1.
12. US Department of Defense, 'Department of Defense Briefing by Gen. Townsend via telephone from Baghdad, Iraq', *US Government, Department of Defense* (28 March 2017), www.defense.gov/News/Transcripts/Transcript/Article/1133033/department-of-defense-briefing-by-gen-townsend-via-telephone-from-baghdad-iraq/ (accessed 15 August 2023).
13. Airwars, 'Airwars Assessment CS598', *Airwars* (2017), https://airwars.org/civilian-casualties/?country=syria&belligerent=coalition&search=cs598 (accessed 15 August 2023).
14. Airwars, 'Airwars Assessment CS598'.
15. Erin Bijl, Wilmoet Wels and Wilbert Van der Zeijden (eds), *On Civilian Harm* (Utrecht: PAX, Protection of Civilians, 2021), p. 161.
16. Drone Wars UK, 'Interview of Air Marshall Greg Bagwell by Chris Cole' (2018), https://dronewars.net/interview-of-air-marshall-greg-bagwell-drone-wars-uk/ (accessed 15 August 2023).

17 Mark Lattimer, 'Interview on the Use of Armed Drones' (3 July 2023).
18 A. MacIntyre, 'The Nature of Virtues', in J. P. Sterba (ed.), *Ethics: The Big Questions* (Hoboken NJ: Wiley-Blackwell, 2009), pp. 371–89.
19 Bernard Williams, 'The Idea of Equality', in R. E. Goodin and P. Pettit (eds), *Contemporary Political Philosophy* (Oxford: Blackwell Publishers Ltd, 1997), pp. 465–75.
20 Richard Rorty, 'Human Rights, Rationality and Sentimentality', *Politike Ljudskih Prava/Politics of Human Rights*, 3/4 (1995), 1/2 (1996) (Belgrade: Beogradski Krug/Belgrade Circle, 1996), 41–57.
21 Isabelle Jones, 'UN Secretary-General Calls for New International Law to Regulate and Prohibit Killer Robots by 2026', *Stop Killer Robots* (20 July 2023), *www.stopkillerrobots.org/news/un-secretary-general-calls-for-new-international-law-to-regulate-and-prohibit-killer-robots-by-2026/* (accessed 15 August 2023).

Select Bibliography

Abelow, Benjamin, *How the West Brought War to Ukraine* (Great Barrington MA: Siland Press, 2022).
Albrecht, U., D. Ernst, P. Lock and H. Wulf, 'Militarization, Arms Transfer and Arms Production in Peripheral Countries', *Journal of Peace Research*, 3/12 (1975), 195–212.
Amstutz, Mark R., *International Ethics: Concepts, Theories, and Cases in Global Politics*, 4th edn (Lanham MD: Rowman and Littlefield Publishers, 2013).
Aron, Raymond, *Peace & War: A Theory of International Relations* (New Brunswick: Transaction Publishers, 2009).
Bacevich, Andrew J., *The New American Militarism: How Americans are Seduced by War* (Oxford: Oxford University Press, 2013).
Beier, J. M., 'Outsmarting Technologies; Rhetoric, Revolutions in Military Affairs, and the Social Depth of Warfare', *International Politics*, 43/2 (2006), 266–80.
Ben-Eliezer, Uri, *The Making of Israeli Militarism* (Bloomington IN: Indiana University Press, 1998).
Berghahn, Volker R., *Militarism: The History of an International Debate 1861–1979* (Cambridge: Cambridge University Press, 1981).
Best, Shaun, 'Zygmunt Bauman: On What it Means to be Included', *Power and Education*, 8/2 (2016), 124–39, https://doi.org/10.1177/1757743816649197 (accessed 15 August 2023).
Bijl, E., W. Wels and W. Van der Zeijden (eds), *On Civilian Harm* (Utrecht: PAX, Protection of Civilians, 2021).
Blum, William, *Rogue State: A Guide to the World's Only Superpower*, 3rd edn (London: Zed, 2006).
Boyle, J., 'The Necessity of "Right Intent" for Justifiably Waging War', in A. F. Lang Jr, C. O'Driscoll and J. Williams (eds), *Just War: Authority, Tradition, and Practice* (Washington DC: Georgetown University Press, 2013), pp. 181–96.
Bramson, L., and G. W. Goethals (eds), *War* (New York: Basic Books, 1964).
Brunstetter, D. R., and Jean-Vincent Holeindre (eds), *The Ethics of War and Peace Revisited: Moral Challenges in an Era of Contested and Fragmented Sovereignty* (Washington DC: Georgetown University Press, 2018).

Burgos, R. A., 'Pushing the Easy Button: Special Operations Forces, International Security, and the Use of Force', *Special Operations Journal*, 4/2 (2018), 109–28.

Calhoun, Laurie, *We Kill Because We Can* (London: Zed Books, 2015).

Ceadel, Martin, *Thinking about Peace and War* (Oxford: Oxford University Press, 1987).

Chamayou, Gregoire, *Drone Theory* (New York: Penguin Books, 2015).

Clark, John W., 'Remote Control in Hostile Environments', *New Scientist*, 22/389 (1964), 300–3.

Clausewitz, Carl Von, *On War* (London: Penguin, 1982).

Coates, A. J., *The Ethics of War* (Manchester: Manchester University Press, 2016).

Cockburn, Andrew, *Kill Chain: Drones and the Rise of High-Tech Assassins* (London: Verso, 2016).

Cohen, M., T. Nagel and T. Scanlon (eds), *War and Moral Responsibility* (Princeton NJ: Princeton University Press, 1974).

Cox, R. W., 'Social Forces, States and World Orders: Beyond International Relations Theory', *Millennium: Journal of International Studies*, 10/2 (1981), 126–55.

Crawford, N. C., 'Blood and Treasure: United States Budgetary Costs and Human Costs of 20 Years of War in Iraq and Syria, 2003–2023', *Watson Institute International & Public Affairs* (2013), 1–27.

Der Derian, James, *Virtuous War*, 2nd edn (New York: Routledge, 2009).

De Roose, F., 'Self-defence and National Defence', *Journal of Applied Philosophy*, 7/2 (1990), 159–68.

Dodds, Klaus, *Global Geopolitics: A Critical Introduction* (London: Pearson Education, 2005).

Dorrien, Gary, *Imperial Designs: Neoconservatism and the New Pax Americana* (New York: Routledge, 2004).

Dowler, L., and J. Sharp, 'A Feminist Geopolitics?', *Space and Polity*, 5/3 (2001), 165–76.

Doyle, M. W., 'Kant, Liberal Legacies and Foreign Affairs', *Philosophy and Public Affairs*, 12/4 (1983), 323–53.

Dreze, J., 'Militarism, Development and Democracy', *Economic and Political Weekly*, 35/14 (2000), 1171–83.

Dunne, T., M. Kurki and S. Smith (eds), *International Relations Theories, Discipline and Diversity*, 4th edn (Oxford: Oxford University Press, 2016).

Durbin, E. F. M., and J. Bowlby, 'Personal Aggressiveness and War', in L. Bramson and G. W. Goethals (eds), *War* (New York: Basic Books, 1964), pp. 81–103.

Eastwood J., 'Rethinking Militarism as Ideology: The Critique of Violence After Security', *Security Dialogue*, 49/1–2 (2018), 44–56.

Eastwood, James, *Ethics as a Weapon of War: Militarism and Morality in Israel* (Cambridge: Cambridge University Press, 2020).
Eide, A., and Thee, M. (eds), *Problems of Contemporary Militarism* (London: Croom Helm, 1980).
Enloe, Cynthia, *Does Khaki Become You? The Militarization of Women's Lives* (London: Pluto Press, 1988).
Enloe, Cynthia, *Globalization and Militarism: Feminists Make the Link* (Lanham MD: Rowman & Littlefield, 2007).
Evans, J., and L. R. Ward (eds), *The Social and Political Philosophy of Jacques Maritain* (London: Geoffrey Bles, 1956).
Feinstein, Andrew, *The Shadow World: Inside the Global Arms Trade* (London: Penguin Books, 2012).
Fouskas, Vassilis, and Bulent Gokay, *The New American Imperialism: Bush's War on Terror and Blood for Oil* (Westport CT: Praeger Security International, 2005).
Freedman, Lawrence, *The Future of War* (London: Penguin Random House, 2017).
Fukuyama, F., 'The End of History', *National Interest*, 16 (1989), 3–18.
Fukuyama, F., 'After Neoconservatism', in D. Skidmore (ed.), *Paradoxes of Power: US Foreign Policy in a Changing World* (New York: Routledge, 2007).
Fussell, Paul, *Wartime: Understanding and Behaviour in the Second World War* (New York: Oxford University Press, 1989).
Gani, J. K., 'Racial Militarism and Civilizational Anxiety at the Imperial Encounter: From Metropole to the Postcolonial State', *Security Dialogue*, 52/6 (2021), 546–66.
Goldich, R. L., 'American Military Culture from Colony to Empire', *Daedalus*, 140/3 (2011), 58–74.
Goodin R. E., and P. Pettit (eds), *Contemporary Political Philosophy* (Oxford: Blackwell Publishers, 1997).
Grayling, A.C., *Among the Dead Cities* (London: Bloomsbury, 2006).
Greene, J. D., B. R. Sommerville, L. E. Nystrom, J. M. Darley and J. D. Cohen, 'An fMRI Investigation of Emotional Engagement in Moral Judgment', *Science*, 293/5537 (2001), 2105–8.
Gregory, D., 'War and Peace', *Transactions of the Institute of British Geographers*, 35/2 (2010), 154–86.
Grossman, Dave, *On Killing: The Psychological Cost of Learning to Kill in War and Society* (New York: Back Bay Books, 2009).
Hambling, David, *Swarm Troopers: How Small Drones will Conquer the World* (Venice FL: Archangel Ink, 2015).
Hamourtziadou, Lily, *Body Count: The War on Terror and Civilian Deaths in Iraq* (Bristol: Bristol University Press, 2021).

Hamourtziadou, L., and J. Jackson, 'Winning Wars: The Myths and Triumphs of Technology', *The Journal of Global Faultlines*, 6/2 (2019), 127–38.

Hanson, Victor Davis, *Father of Us All: War and History, Ancient and Modern* (New York: Bloomsbury, 2010).

Harding, Luke, *Invasion: Russia's Bloody War and Ukraine's Fight for Survival* (London: Guardian Faber, 2022).

Henderson, E., 'Hidden in Plain Sight: Racism in International Relations Theory', in A. Anievas, N. Manchanda and R. Shilliam (eds), *Race and Racism in International Relations: Confronting the Global Colour Line* (Abingdon: Routledge, 2014), pp. 19–45.

Ikenberry, G. J., and C. A. Kupchan, 'Socialization and Hegemonic Power', *International Organization*, 44/3 (1990), 283–315.

Iraq Body Count, 'A Dossier of Civilian Casualties 2003–2005' (2005), www.iraqbodycount.org/analysis/reference/pdf/a_dossier_of_civilian_casualties_2003-2005.pdf (accessed 15 August 2023).

James, W., 'The Moral Equivalent of War', in L. Bramson and G. W. Goethals (eds), *War* (New York, London: Basic Books, 1964), pp. 21–31.

Johnson, Chalmers, *The Sorrows of Empire: Militarism, Secrecy, and the End of the Republic* (New York: Metropolitan Books, 2004).

Jordan, D., 'Air and Space Warfare', in D. Jordan, J. D. Kiras, D. J. Lonsdale, I. Speller, C. Tuck and C. D. Walton (eds), *Understanding Modern Warfare* (Cambridge: Cambridge University Press, 2008), pp. 179–223.

Kant, I., *Political Writings* (Cambridge: Cambridge University Press, 2007).

Kasher, A., and A. Yadlin, 'Military Ethics of Fighting Terror: An Israeli Perspective', *Journal of Military Ethics*, 4/1 (2005), 3–32.

Kaufman, S., R. Little and W. C. Wohlforth, *The Balance of Power in World History* (New York: Palgrave Macmillan, 2007).

Kennan, G. F., 'Morality and Foreign Policy', *Foreign Affairs*, 64/2 (1985–6), 205–18.

Kilcullen, David, *The Dragons and the Snakes: How the Rest Learned to Fight the West* (London: Hurst, 2022).

Kiras, J. D., *Special Operations and Strategy: From World War II to the War on Terrorism* (Abingdon: Routledge, 2006).

Koenigsberg, R., *Nations have the Right to Kill: Hitler, the Holocaust, and War* (New York: Library of Social Science, 2009).

Kristol, W., and R. Kagan (eds), *Present Dangers: Crisis and Opportunity in American Foreign and Defense Policy* (New York: Encounter Books, 2000).

Lang Jr, A. F., C. O'Driscoll and J. Williams (eds), *Just War: Authority, Tradition, and Practice* (Washington DC: Georgetown University Press, 2013).

Lebow, R. N., 'Classical Realism', in T. Dunne, M. Kurki and S. Smith (eds), *International Relations Theories, Discipline and Diversity*, 4th edn (Oxford: Oxford University Press, 2016), pp. 34–50.

Lee, Peter, *Reaper Force* (London: John Blake Publishing, 2018).

Lefever, Ernest W., *Ethics and World Politics* (Washington DC: Ethics and Public Policy Center, 1988).

Lieblich, E., 'The Facilitative Function of *Jus in Bello*', *The European Journal of International Law*, 30/1 (2019), 321–40.

Lovin, R. W., *Reinhold Niebuhr and Christian Realism* (Cambridge: Cambridge University Press, 1995).

MacIntyre, A., 'The Nature of Virtues', in J. P. Sterba (ed.), *Ethics: The Big Questions* (Hoboken NJ: Wiley-Blackwell, 2009), pp. 371–89.

MacKenzie, D., 'Militarism and Socialist Theory', *Capital and Class*, 7/1 (1983), 33–73.

Maguen, S., D. D. Luxton, N. A. Skopp, G. A. Gahm, M. A. Reger, T. J. Metzler and C. R. Marmar, 'Killing in Combat, Mental Health Symptoms, and Suicidal Ideation in Iraq War Veterans', *Journal of Anxiety Disorders*, 25/4 (2011), 563–67.

Mann, M., 'Authoritarian and Liberal Militarism: A Contribution from Comparative and Historical Sociology', in S. Smith, K. Booth and M. Zalewski (eds), *International Theory: Positivism and Beyond* (Cambridge: Cambridge University Press, 1996), pp. 221–39.

Marx, Anthony W., *Faith in Nation: Exclusionary Origins of Nationalism* (New York: Oxford University Press, 2005).

Mastanduno, M., 'Hegemonic Order, September 11, and the Consequences of the Bush Revolution', *International Relations of the Asia Pacific*, 5 (2005), 177–96.

May, Larry, *War Crimes and Just War* (Cambridge: Cambridge University Press, 2007).

Mearsheimer, John, *The Tragedy of Great Power Politics* (New York: Norton, 2001).

Middlebrook, Martin, *The Battle of Hamburg: The Firestorm Raid* (London: Allen Lane, 1980).

Milgram, Stanley, *Obedience to Authority: An Experimental View* (New York: Harper and Row, 1974).

Mills, C. Wright, *The Power Elite* (Oxford: Oxford University Press, 1956).

Montgomery, Bernard Law, *Memoirs* (London: Collins, 1958).

Morgenthau, Hans, *Scientific Man vs. Power Politics* (London: Latimer House, 1947).

Morgenthau, Hans, *Decline of Domestic Politics* (Chicago IL: University of Chicago Press, 1958).

Morgenthau, Hans, *Politics Among Nations: The Struggle for Power and Peace* (New York: Alfred A. Knopf, 1973).
Moyn, Samuel, *Humane: How the United States Abandoned Peace and Reinvented War* (London: Verso, 2022).
Munkler, Herfried, *The New Wars* (Cambridge: Polity, 2004).
Neillands, Robin, *The Bomber War: The Allied Air Offensive Against Germany* (New York: Overlook Press, 2001).
Orend, Brian, *War and Political Theory* (Cambridge: Polity, 2019).
Porch, D., 'Expendable Soldiers', *Small Wars & Insurgencies*, 25/3 (2014), 696–716.
Prados, J., 'The Continuing Quandary of Covert Operations', *Journal of National Security Law and Policy*, 5/2 (2012), 359–72.
Primoratz, I. (ed), *Terror from the Sky: The Bombing of German Cities in World War II* (Oxford, New York: Berghahn Books, 2014).
Puri, Samir, *Russia's Road to War with Ukraine* (London: Biteback Publishing, 2022).
Rees, Laurence, *Auschwitz: The Nazis and the 'Final Solution'* (London: Penguin Random House, 2005).
Robins, K., and L. Levidow, 'Socializing the Cyborg Self: The Gulf War and Beyond', in C. H. Gray (ed.), *The Cyborg Handbook* (New York: Routledge, 1995), pp. 119–25.
Rogers, P., 'The Triple Paradigm Crisis: Economy, Environment, and Security', *The Journal of Global Faultlines*, 6/2 (2019), 139–49, *www.scienceopen.com/hosted-document?doi=10.13169/jglobfaul.6.2.0139* (accessed 15 August 2023).
Rogers, Paul, *Losing Control: Global Security in the Twenty-First Century*, 4th edn (London: Pluto Press, 2021).
Ross, A. L., 'Dimensions of Militarization in the Third World', *Armed Forces and Society*, 13/4 (1987), 561–78.
Rutchick, A. M., R. M. McManus, D. M. Barth, R. J. Youmans, A. T. Ainsworth and J. H. Goukassian, 'Technologically Facilitated Remoteness Increases Killing Behavior', *Journal of Experimental Social Psychology*, 73 (2017), 147–50.
Sareen, J., B. J. Cox, T. O. Afifi, M. B. Stein, S. L. Belik, A. Meadows and J. G. Gordon, 'Combat and Peacekeeping Operations in Relation to Prevalence of Mental Disorders and Perceived Need for Mental Health Care', *Arch Gen Psychiatry*, 64/7 (2007), 843–52.
Scott, Peter Dale, *The Road to 9/11: Wealth, Empire and the Future of America* (Los Angeles CA: University of California Press, 2007).
Sebald, W. G., *On the Natural History of Destruction* (London: Random House, 2003).
Shaw, Martin, *Dialectics of War: An Essay in the Social Theory of Total War and Peace* (London: Pluto Press, 1988).

Shaw, Martin, *Post-Military Society: Militarism, Demilitarisation, and War at the End of the Twentieth Century* (Philadelphia PA: Temple University Press, 1991).
Shaw, Martin, *War and Genocide: Organized Killing in Modern Society* (Cambridge: Polity, 2003).
Shaw, Martin, *The New Western Way of War: Risk-Transfer War and its Crisis in Iraq* (Cambridge: Polity, 2005).
Sigwart, H. J., 'The Logic of Legitimacy: Ethics in Political Realism', *The Review of Politics*, 75/3 (2013), 407–32.
Singer, P. W., *Wired for War* (New York: Penguin Books, 2010).
Stavrianakis, A., and J. Selby (eds), *Militarism and International Relations: Political Economy, Security, Theory* (London: Routledge, 2013).
Steel, Ronald, *Pax Americana* (New York: Viking, 1967).
Sterba, J. P. (ed.), *Ethics: The Big Questions* (Hoboken NJ: Wiley-Blackwell, 2009).
Stohl, Rachel, and Suzette Grillot, *The International Arms Trade* (Cambridge: Polity, 2009).
Strawser, B. J., 'Moral Predators: The Duty to Employ Uninhabited Aerial Vehicles', *Journal of Military Ethics*, 9/4 (2010), 342–68.
Suganami, Hidemi, *On the Causes of War* (Oxford: Clarendon Press, 1996).
Taddeo, M., 'Trusting Digital Technologies Correctly', *Minds and Machines*, 27/4 (2017), 565–8.
Talbot, K., 'The Real Reasons for War in Yugoslavia: Backing Up Globalization with Military Might', *Social Justice*, 27/4 (2000), 94–116.
Thompson, E. P., 'Notes on Extremism: The Last Stage of Civilization', in *Exterminism and Cold War*, New Left Review (London: Verso, 1982), pp. 1–33.
Tritle, L., 'Warfare in Herodotus', in C. Dewald and J. Marincola (eds), *The Cambridge Companion to Herodotus* (Cambridge: Cambridge University Press, 2006), pp. 209–23.
Vagts, Alfred, *A History of Militarism: Civilian and Military*, 2nd edn (New York: Meridian Books, 1959).
Vagts, Alfred, *A History of Militarism: Romance and Reality of a Profession* (New York: W. W. Norton and Co., 1973).
Van der Pijl, Kees, *The Making of an Atlantic Ruling Class* (New York, London: Verso, 1984).
Vilmer, J.-B. J., 'France and the American Drone Precedent', in R. Daniel Brunstetter and J.-V. Holeindre (eds), *The Ethics of War and Peace Revisited: Moral Challenges in an Era of Contested and Fragmented Sovereignty* (Washington DC: Georgetown University Press, 2018), pp. 97–116.

Waldman, Thomas, *Vicarious Warfare: American Strategy and the Illusion of War on the Cheap* (Bristol: Bristol University Press, 2021).

Wall, T., and T. Monahan, 'Surveillance and Violence for Afar: The Politics of Drones and Liminal Security-Scapes', *Theoretical Criminology*, 15/3 (2011), 239–54.

Walzer, Michael, *Just and Unjust Wars*, 3rd edn (New York: Basic Books, 2000).

Weil, Simone, *Formative Writings* (New York: Routledge, 1999).

Wight, Martin, *Power Politics* (London: Royal Institute of International Affairs, 1946).

Williams, B., 'The Idea of Equality', in R. E. Goodin and P. Pettit (eds), *Contemporary Political Philosophy* (Oxford: Blackwell Publishers Ltd, 1997), pp. 465–75.

Williams, Bernard, *Ethics and the Limits of Philosophy* (London: Routledge, 2006).

Woodward, Bob, *Obama's Wars* (New York: Simon and Schuster, 2010).

Yenne, Bill, *Attack of the Drones: A History of Unmanned Aerial Combat* (London: Zenith Press, 2004).

Index

A
Afghan civilians 126
air power 6, 33–4, 39, 136
air raids 35, 37, 43, 100, 158
air strikes 39, 102, 107–9, 111
American exceptionalism 119
attacks, swarm 165
Auschwitz 46, 48–9, 159
Autonomy 14, 68–9, 72–3, 90, 119

B
battlefield 11, 66, 110, 117, 135, 141–2, 170
Bayraktar TB2 military drones 162
Blitz 6, 35, 41
Bomber War 34, 42, 45, 57
bombings, atomic 37–8, 42
British militarism 44
Bush doctrine of preventive war 116

C
Camp Speicher 130
casualty recording 171–2, 199
China's combat drones 158
Christian realists 18
civilian casualties 63, 91, 122, 128, 133
Clausewitz, Carl Von 15–16, 27–8, 42, 45, 77, 133
Compassion 25, 86
computerised postmodern warfare 81
Coventry 35
Crusading 23–4, 44
cyberattacks 2, 115

D
darkness, moral 115
database of Iraq Body Count 106
death camps 34, 46, 53, 92, 176
delegation 7, 47, 53, 111, 125–6, 154, 161, 166
destruction, mass 32, 58, 119
Deus ex machina 27, 125
distance, safe 41, 111, 115
dominance 19, 97–8, 120
drone boats 143, 147
drone operators 72, 81, 155
drones
  commercial 8, 136
  kamikaze 3, 143, 146
  submarine 64
  subterranean 64
  suicide 143
  terrestrial 64
drone technology 166
drone war in Pakistan 77

E
emotion and moral value 83
emotions 72–3, 79, 83
executions 94, 101–2, 106, 115, 118, 130–1

F
First World War 17, 21, 33, 59, 118–19
Fu-Go 60
Fukuyama, Francis 95

G
gas chambers 6, 47–8, 52–3
Geneva Conventions 54, 94
Geneva Conventions and international humanitarian law 58
Genocide 12, 21, 32, 54, 63, 131, 159

German Cities 35–6, 132
German Luftwaffe 34
global arms trade 8, 156, 161
global security 157
global war business 165
glorification 23, 44, 119
GPS (global positioning systems) 3, 66, 69

H
hegemonic power 153
heroism 65, 76–8, 126–7, 174
Holocaust 48
human security 4, 8

I
Imperialism 23–4, 44, 97, 119, 124, 154, 161
incendiary bombs 37, 60
international arms trade 156
international ethics 12
International Peace Conference 38
invasion and occupation of Iraq 8, 95, 109, 126–7
Iran 86, 91, 96–7, 115, 161, 164
Iranian and Turkish drones 158
Iraq and Afghanistan 78, 110–11
Iraq Body Count 98, 101–2, 104, 106, 129–30, 171
Iraqi civilians 98, 107, 128, 130, 160
Iraq's casualties 171
ISIS fighters 130, 172

J
Jus in bello 28, 31, 54, 58, 131, 133, 136, 144
Just War Theory and International Humanitarian Law 32

K
Kabul airport in Afghanistan 126
Kamikaze Switchblade 8, 142–3, 159
Kant, Immanuel 77, 122

Kherson region 143–4
Kill Chain 101
killing, cold-blooded 74
Kosovo 61–4, 90, 122

L
laws of armed conflict 29, 31–2, 173
LITTLE BOY 93
Living under drones report 77
loitering munitions 3, 8, 142–4, 168

M
martial virtues 79
mass graves 106–7, 131
Middle East 7, 86, 96–7, 141, 157, 165
militarism, racial 53–4
military ethics 22, 73
military ethos 73, 76, 78
military technology 116, 119
military valour 79
missiles, long-range 142, 159
moral buffering 82
Moral disengagement 110
morality in war 43
moral soiling 71
munitions, precision-guided 98, 129

N
national anthems 79, 81, 146
national defence 8, 136, 139, 159
nationalism 16, 75, 79
nation of heroes 146
NATO 62, 139, 149, 151–3, 156, 160–1
NATO military exercises 150
NATO strikes in Yugoslavia 90
NATO-Ukraine-Russia war 154
Necroethics 70
Nuremberg principles 12

O
Obama, Barack 102, 115, 117, 121
obligations 17, 25, 29, 43, 87

Operation Gomorrah 36
Operation Iraqi Freedom 98, 104

P
Pax Americana 110, 119
peace, perpetual 122
pilots 33, 41, 58, 61–2, 67–8
Predator 27, 61, 64, 76, 94, 98, 104, 158
preemptive strikes 30, 130
profit 8, 156–7, 161, 166
Projecting power 75
proper authority 29–1, 43, 148
proportionality 29–30, 32, 63, 75, 85–6, 132–3, 147, 169
proxies 2, 112, 114, 159, 166
Putin 139, 152, 160

Q
al-Qaeda 77, 93, 95, 164

R
RAF bombing of German civilians 35
realism 13–16, 18–19, 53, 55, 82, 91, 121–4, 126, 145, 160
Reaper drones 84
Remote-control killing 94
remoteness 50–1, 111, 115, 161
reprisals 32, 35, 43, 129–30
robots 66–7, 72–3
    killer 167–9, 175
Royal Air Force (RAF) 34, 40
Russian and Ukrainian casualties 150
Russian drone attacks 143, 155

S
Sacrifice 44, 73, 76, 79–80, 120
Security Council 30, 62
Shahed drones 145
SIPRI (Stockholm International Peace Research Institute) 157–8

Sonderkommandos 46–8
Switchblade drone 147

T
Taliban 77, 95, 97, 113, 129
terrorising 39, 141, 159
terrorism, global 94, 121
terror of the air 40

U
UAS (unmanned-aircraft system) 142, 151, 161
UAVs (unmanned aerial vehicles) 56, 61, 93, 132, 157–8, 162, 166
Ukraine's Drone Wars 144, 160
Ukrainian civilians 143, 145, 148
Ukrainian quadcopters 141
US drone strikes hit Yemen 104

V
virtuous war 4, 26–7, 125

W
war crimes 12, 32, 56, 73, 82, 128
war ethics 24–5, 43, 54, 171
War on Terror in Afghanistan 142
war on terrorism 95–6
weapons
    new 58, 70, 171
    remote 141, 159, 167
Wing Loong II drones 164

Y
Yazidis 131

Z
Zelenskyy, Volodymyr 140, 145, 148, 152
Zyklon 48–49